WAR and the Red ✚ Cross

WAR and the Red ✚ Cross

THE UNSPOKEN MISSION

Nicholas O. Berry

St. Martin's Press
New York

WAR AND THE RED CROSS: THE UNSPOKEN MISSION

 For information, address St. Martin's Press, 175 Fifth Avenue, New York, N.Y. 10010.

ISBN 0–312–16517–X

Library of Congress Cataloging-in-Publication Data

Berry, Nicholas O.
War and the Red Cross : the unspoken mission / Nicholas O. Berry.
p. cm.
Includes bibliographical references and index.
ISBN 0–312–16517–X
1. Red Cross. 2. Humanitarian assistance. 3. Pacific settlement of international disputes. 4. Peaceful change (International relations) 5. War—Termination. I. Title.
HV568.B47 1997
361.7'7—dc21 96-48926
CIP

Design by Harry Katz, Digital Type & Design

First edition: June 1997
10 9 8 7 6 5 4 3 2 1

Nine delegates of the International Committee of the Red Cross were murdered while this book was in preparation. This dedication is a memorial to their bravery.

Died in Burundi, June 4, 1996

Cédric Martin
Reto Neuenschwander
Juan Pastor Ruffino

Died in Chechnya, December 17, 1996

Nancy Malloy
Fernanda Calado
Sheryl Thayer
Gunnhild Myklebust
Ingeborg Foss
Johan J. Elkerbout

Contents

Preface

This is a story about the world's leading war-relief organization and its unspoken mission to undermine and resolve civil wars. It is also an analysis—illustrated by six case studies—of the changing nature of these wars.

At times, the subject seemed overwhelming. And, as the dedication indicates, I was dealing with sensitive information that could affect the lives of Red Cross workers who accept grave risks in order to bring protection and relief to the victims of war. More than once I thought of abandoning the project. The story, however, needed telling. I hope it adds a small voice to the shouts of those who labor against the terrible institution called war.

There were always people who helped. Ursinus College was generous in its support, as was its library staff, especially Judith E. Fryer and David H. Mill. My colleagues, Steven Hood, Paul Stern, Gerard Fitzpatrick, Houghton Kane, and William Akin gave advice and encouragement. The graphics were expertly designed by Paul Bashus. Melissa Grafe compiled the index. I learned some essential legal and political-military information at seminars, for which I thank John Norton Moore, Director of the Center for National Security Law at the University of Virginia School of Law; Scott L. Silliman, Executive Director of the Center on Law, Ethics and National Security at the Duke University School of Law; and Pamela Aall, Deputy Director of the Education and Training Program at the United States Institute of Peace. My thanks to Colonel Charles J. Dunlap, Jr. at the Air Force Judge Advocate's Office for helping me understand the practical problems involved with peacekeeping operations, especially those he experienced in Somalia. Nik Gowing, diplomatic editor for ITN's Channel Four News in London, shared his thoughts on the impact of real-time television coverage of wars on foreign policy decisions.

All of us who write about humanitarian action in today's wars owe much to the pioneering studies of Larry Minear and Thomas G. Weiss, both at the Thomas J. Watson Jr. Institute for International Studies at Brown University, and to David P. Forsythe at the Political Science Department at the University of Nebraska at Lincoln.

International Committee of the Red Cross officials, delegates, and staff were most forthcoming with documents, photographs, and papers. They welcomed me into their offices in Geneva, New York, and Guatemala City. They patiently answered my often indelicate questions, and a number of them kindly read portions of the manuscript to correct errors of fact or interpretation. Any errors that remain, of course, are my own. Some ICRC officials thought I exaggerated the ICRC's unspoken mission. Others did not. Some, quite frankly, were displeased to have the politics of the ICRC out in public. Others were not. All were helpful, and my deep appreciation goes to ICRC President Cornelio Sommaruga, André Pasquier, Daniel Augstburger, Patrick Zahnd, Graziella De Vecchi, Laurent Burkhalter, François Bugnion, Olivier Dürr, Sylvie Junod, Jean-Paul Fallet, Carlo von Flüe, Kim Gordon-Bates, Tony Burgener, and Alan Dorsey.

Finally, my wife Janet has again proven to be an indispensable editor for shaping-up my all too frequently highly original prose. She made the work of the editors at St. Martin's Press—Michael Flamini, Elizabeth Paukstis, Wendy Kraus, and Diana Gavales—less of a challenge.

NICHOLAS O. BERRY

WAR and the Red ✚ Cross

Introduction

Happenstance can plant a small seed that grows and sometimes flowers.

During a luncheon at a 1994 conference in Washington sponsored by the United States Institute of Peace, I happened to sit next to Mr. André Pasquier, special advisor to the president of the International Committee of the Red Cross (ICRC). The subject of the conference was "Managing Chaos," subtitled: "Coping with International Conflict in the 21st Century." Mr. Pasquier was one of a great number of representatives of non-governmental organizations (NGOs) at the conference.* These voices from Africare, World Vision, InterAction (an association of NGOs), Refugees International, International Rescue Committee, and other NGOs involved in relief operations spoke about the suffering in Rwanda, Somalia, Bosnia, and Angola and their role in meeting the desperate needs of the victims of armed conflicts. Speakers from the U.S. government and the UN praised the NGOs for their efforts. Yet, any reference to the politics of these wars seemed out of bounds whenever humanitarian relief was discussed.

I asked Mr. Pasquier why politics was poison for most NGOs involved in international humanitarian relief operations. His answer was persuasive, but it turned out to be only half the story. My purpose here is to present both halves of the story. No organization that is heavily involved in wars, I thought, can do so without having a major impact on the politics of the wars. There must be a reason the ICRC does not want to talk about it openly.

And there is.

Only by avoiding politics, Mr. Pasquier told me, can the ICRC fulfill its mission of assisting the victims of war. The Fundamental Principles and the Code of Conduct the ICRC follows stipulate absolute political neutrality and impartiality in its involvement in both international and internal wars. Avoiding politics establishes access to *all sides* in war-torn areas, the ICRC's "top priority." Supporting one side would upset the power relations between the warring parties and thus preclude any access to the other side. Avoiding politics helps protect humanitarian workers and furthers their relief efforts by not making them enemies (and thus targets) of any of the warring parties. The ICRC is deeply

---

* The ICRC, a totally Swiss corporation, is technically not an international NGO. Moreover, in comparison to NGOs, the mandate given to the ICRC by the Geneva Conventions endows it with a special status in providing humanitarian assistance in wars.

concerned that recent UN peace enforcement operations under Chapter VII of the Charter, which requires the non-neutral use of military force, taint the ICRC's relief workers with the UN's non-neutral political bias. Avoiding politics also facilitates donations and cooperation from governments by guaranteeing a humanitarian, not political, role. And it advances efforts to disseminate international humanitarian law by the ICRC's impartially following international law itself.

Paradoxically, only this neutral and impartial non-political stance by the legally designated authority in war relief allows it to play positive political roles. That is, the *effects* of its non-political stance tend to corrupt war as an instrument of state policy and discourage its use in resolving conflicts. It took me six months to come to this conclusion and another six months to prove that the ICRC had indeed taken on this new, but unpublicized, mission.

The larger aspect of this study traces the changing nature of modern war. As a human institution, war is going the way of dueling, slavery, and the subordination of women. War is becoming obsolete. Studying the ICRC in action reveals the world's premier war relief agency engaged in a campaign to hasten war's obsolescence.

The ICRC has been involved in virtually every war since the latter third of the nineteenth century. For the first time, the ICRC, other NGOs, and intergovernmental organizations (IGOs) such as the UN are having a tremendous impact on war itself. These third parties, the ICRC realized soon after the end of the Cold War, have the capability of undermining war as a method of resolving civil conflicts. Today's wars are internal, small, intolerably brutal, and, most importantly, *highly vulnerable to outside interference.* In recognition of this monumental discovery, the ICRC led the entire International Red Cross and Red Crescent Movement to (implicitly) declare war on war at its 26th International Conference in Geneva in December 1995. Its hostility to today's brutal civil wars was barely disguised. The ICRC's main mission, to protect the victims of war, is now joined by a second: hastening the erosion of the efficacy of war. This is the other half of the story.

The ICRC's non-political stance in war maximizes the ICRC's presence in war zones, which is the prerequisite for its unspoken mission of softening and ending civil wars. This is why this new mission is not publicized or even presented as policy, because to advertise or even formalize this new mission would sully the ICRC's overt independent, impartial, and neutral humanitarian principles and operations. Getting ICRC officials to frankly discuss this second unspoken mission has proved to be difficult (but not impossible) for obvious reasons. Loudly proclaiming its new mission would hinder the ICRC from accomplishing its main mission of war relief. Warring parties, while expending blood and treasure, would hesitate to invite into the fray a third party that would hamper their efforts to win the war. On the other hand, proclaiming and doing the stated mission has the effect of accomplishing its new mission! Ergo, the ICRC keeps quiet about its new anti-war agenda.

Specifically, the political effects of ICRC and other "non-political" NGO relief operations undermine the institution of war in the following ways:

1. *They induce governments and UN agencies to provide massive assistance to the victims of war and even to intervene as peacemakers, peacekeepers, or peace enforcers.* What is imprecisely called the "CNN factor" is often first generated by the actions of NGO officials publicizing horrible pictures and stories of massive human wartime suffering. The ICRC seeks to create an international concern about wars by using its extensive, private communication channels and its media connections. When the UN or another third party then intervenes in wars, the outcome is taken out of the hands of the warring parties to a great extent. Warring parties lose control and find themselves pressed to settle the conflict. "Humanitarian action must be in parallel with political actions," declared ICRC President Cornelio Sommaruga at the 1995 International Conference. Together, humanitarian agencies and peacekeepers create a full range of interference with the warring parties' use of force.

2. *They constrain the level of violence by monitoring and thus deterring many instances of brutality.* Intrepid ICRC workers put their bodies at risk in war zones. Their access to the media implicitly threatens any warring party with disclosure if it commits war crimes. This limits the power of combat, which, after all, is about killing people.

3. *They publicize abuses by any warring party, thereby getting governments in the international community to exert corrective pressure and sanctions on that abusive party.* An alignment of outside forces is created. UN or NATO peacekeepers can be influenced by the ICRC to put a stop to the criminal behavior of a warring party, or at least to attempt to reduce it.

4. *They help stalemate wars by protecting and sustaining the populations of all warring parties, thus making a diplomatic settlement more probable than the total victory of one side. Internal wars are fought over civilian loyalty and the ability of civilians to sustain their fighters in the field.* To provide food, health care, shelter, and protection to *all* civilians, therefore, weakens the foundation of these wars by insulating civilians from the warring parties and by building antiwar sentiments among the population. The ICRC interferes with the human sustenance of war and thereby discourages it.

5. *They mobilize the international community to adopt and comply with international humanitarian law, thus sharply restricting the use of force in combat and the ability of wars to resolve conflicts.* Rules governing armed conflicts are based on the principle

of protecting the civilian population from the effects of war. For example, the banning of inhumane but efficient weapons, such as poison gas and, now, antipersonnel mines, has always been high on the ICRC's agenda.

Where in recent wars have there been decisive victories? The Iran-Iraq War? The second Gulf War? Cambodia? Angola? Mozambique? Somalia? Bosnia? Where has war shown its robust nature, its past glories in resolving great conflicts? I will demonstrate that very few expected or recognized the factors that have helped the ICRC undermine civil wars. International humanitarian law has not interfered that much with the conduct of interstate wars. Political objectives, not humanitarian relief, dominated. Civilians were usually not the prime target of combat, POWs could be sequestered, and the wounded could be treated by large organized armies without affecting the status of combat-ready troops. International wars were large, and their scope tended to overwhelm the presence of NGOs and other third parties. Complying with international humanitarian law did little to inhibit the striving for victory.

Internal wars changed the equation. Now, international humanitarian law really mucks up combat. Civilians are the target because they sustain the warring parties, and how they align themselves determines a war's outcome. POWs are a big burden on the rag-tag local combatants, and the wounded, if cared for, can be back in the war zone within days. And the scope of internal war is small, a fact that effectively enlarges the presence and effects of any third party. Thus, following international humanitarian law and allowing the ICRC, other NGOs, and UN agencies to intervene disrupts the balance of power. In particular, UN intervention injects a powerful force backed by major powers into wars that, compared to interstate wars—with their massed tanks, air forces, and navies—are fought between *weaker* warring parties. This *magnifies* the relative power of the third parties—the UN, the ICRC, and other NGOs.

It is thus now harder to win wars. Stalemates ensue. A brokered peace becomes the way out.

6. They influence the operations of the UN in order to make them consistent with international humanitarian law, thereby reinforcing antiwar peacekeeping, inducing war-crimes tribunals, and making more effective UN-brokered conflict resolution.

7. They create neutral intermediaries between the warring parties for messages, meetings, and exchanges. This serves to breakdown the stereotyping and hatred so necessary for the conduct of war, especially for internal wars.

The ICRC's unspoken new mission is nothing less than the sabotage of war. Its attack on war comes from many directions. It builds coalitions of intervening third parties, interfering with the contest of arms. It propagandizes against the horrors of war, giving war a bad name. It injects a measure of humanity into wars, testifying to a civilized alternative to killing. It protects civilians, the political targets of civil wars. It pressures all political groups to comply with international humanitarian law in order to limit combat and its effects. It encourages mediation. If the ICRC's historic mission, as it proclaims in a recent brochure, "is to mobilize the resources of civil society to come to the assistance of victims of war and to provide humanitarian assistance to vulnerable people worldwide," then is it not logical that the first line of defense to relieve suffering is to *prevent war* by making war an ineffective institution? No wonder the ICRC sublimates its political role.

On the negative side, stalemating wars sometimes prolongs wars—and thus suffering—more so than if one warring party were allowed to move to a quick, decisive victory. Relief supplies can also be used to sustain the warriors. In addition, the availability of humanitarian relief workers sometimes leads to excess "collateral damage" to civilians inflicted by the warring parties, who know that relief workers will soon arrive to clean up the mess.

On balance, however, the "non-political" role of international relief agencies (since their politics is masked by a neutral stance) promotes highly beneficial political effects (undermining the institution of war and promoting negotiated solutions). Any overt partisan political role would destroy virtually all their humanitarian efforts to provide protection and assistance to all the victims of war. Just as important, any overt political role would unmask the ICRC's efforts to undermine civil wars (a task made easier in these wars because they are small and fragile) and open its operations to manipulation by the warring parties as they attempt to counter any interference in their struggle.

International politics used to be the exclusive domain of governments and the international organizations, alliances, and coalitions they formed. No longer. In coping with conflicts in the twenty-first century, NGOs will at times coordinate with governments and their international organizations, at times cooperate with them, and at times oppose their actions. And all the while they will profess an indifference to politics that is sincere but patently deceptive.

Because of the ICRC's global presence and its special legal status in both international and internal wars, its actions will reveal this emerging pattern of international politics and the growing role of NGOs, guided and inspired by the ICRC, in successfully discouraging armed conflicts.

We are witnessing the growing obsolescence of war as a human institution. It is a momentous turn of history.

A final introductory note. Will this study, which for the first time candidly explores the extensive political effects of the world's premier war-relief agency, damage the ability of that agency to provide relief? You would not be reading this if the answer were "yes." ICRC officials in my interviews have made no secret of their intense disapproval of all the current civil wars that violate international humanitarian law. And the ICRC's laudable humanitarian actions speak for themselves. Warring parties need ICRC relief services. It "just happens" that they have a wide array of antiwar consequences, some of which are intended and some that were only revealed after analyzing the experience of the massive, post–Cold War intervention in internal wars. Recent UN intervention has magnified the political effects of NGOs. Above all, the ICRC's role in war relief is part of international humanitarian law, ratified by virtually all states. Who would argue that the wounded should not be nursed, civilians not protected, refugees not sheltered or fed, and missing family members not traced? If this gets in the way of war, then war must accept the limits governments now place upon it. Too bad for war.

1

The Main Mission: The Origins and Development of the Red Cross

Wars make history by creating and destroying states, making and unmaking governments, shifting territorial boundaries, and validating the ideologies of the winners and discrediting those of the losers. Wars are high-risk attempts to settle disputes in decisive ways. Wars test political skills and personal courage. They unify people with the glue of cultural heroes and events, which then become milestones for nations, ethnic groups, and religions. For those in battle, wars can be, in the short run, exhilarating, mainly to the winners.

Something with as many monumental effects as war must have great political value, but are wars a good and pleasing fulfillment of human nature? If the memoirs of warriors are any guide, the answer is no. They express anger, boredom, terror, remorse, and fatigue. Even soldiers on the winning side see destruction and carnage as friends die or suffer physical and mental wounds. In essence, war is a loud and ugly thing.

The Red Cross originally focused on the loud and ugly and sought to bring a measure of humanity or more silence and more beauty to wars, if that is possible to conceive. Initially, the Red Cross accepted wars as an inevitable part of history. Its first mission was to ameliorate the conduct and effects of war. Only after the end of the Cold War, when it became possible, did the Red Cross take on the additional mission of making internal wars dysfunctional.

A BRIEF HISTORY

On June 24, 1859, Henry Dunant, a young Swiss businessman, happened across the aftermath of a great battle and was appalled at the suffering of the wounded. Over 40,000 French and Italian soldiers were killed or wounded that day at Solferino, Italy. The few who cared for those still alive, small military medical staffs and random townspeople, could not provide adequate shelter, medicine, food, and water to those suffering. Thousands died of neglect. Dunant hastily organized more local citizens to care for the wounded, regardless of nationality, using his own resources to buy supplies. He learned firsthand that armies were not prepared to nurse their sick and wounded.[1]

Returning to his home in Geneva and haunted by the experience, Dunant wrote about it in a small book entitled *Un Souvenir de Solferino,* (A Memory of Solferino, 1862). He described the deaths that could easily have been prevented with a little care. "Would it not be possible in time of peace and quiet," Dunant suggested, "to form relief societies for the purpose of having care given to the wounded in wartime by zealous, devoted and thoroughly qualified volunteers?"[2]

The book was widely read. In 1863, it moved a prominent Swiss, Gustave Moynier, president of the Welfare Society of Geneva, to organize a committee with Dunant for the purpose of creating relief societies to care for the tattered human residue of war. With Dunant continually and effectively promoting the need for such societies, Moynier's International and Permanent Committee for the Relief to Wounded Military Personnel saw sufficient interest to call an international convention. Sixteen European governments sent representatives to Geneva in August 1864, and twelve signed the Geneva Convention for the Amelioration of the Condition of the Wounded and Sick of Armies in the Field. It provided for the neutrality and inviolability both of armed forces' medical personnel and of the civilian volunteers who assisted them in the field. It prescribed humane treatment of the wounded, no matter on what side they fought or on what ground they fell. It included preliminary provisions for dealing with prisoners of war. Finally, it created an emblem to distinguish the neutral relief workers (and their equipment) who would aid the wounded strictly as noncombatants. In honor of Geneva and the conference organizers, the red cross on a white background—the reverse colors of the Swiss flag—was chosen.[3]

The Committee almost immediately became commonly known as the Red Cross, and its official name changed to the International Committee of the Red Cross. To become international and permanent and to develop, in Moynier's words, "the principles of humanity," the Committee embarked on three tasks. It had to create a worldwide movement by encouraging the formation of national Red Cross societies and then giving them official recognition. It had to further develop international humanitarian law and then induce more states to adhere to it. Finally, it had to structure the entire Red Cross movement to accomplish the first two tasks as well as its mission of providing relief to those wounded in war.

The Committee's organizing efforts centered first on Europe, where 12 governments had already signed the 1864 Geneva Convention. It would recognize national societies whose governments had signed the Convention and given them official recognition. With no pause, the Committee sought out non-European notables to organize in their native lands. The Ottoman Society, for instance, was recognized in 1868 (and in 1876 adopted the red crescent symbol), the Peruvian Society in 1879, and the Japanese Society in 1886. Societies in most Muslim countries are called the Red Crescent and use that as their symbol. Iran used the name Red Lion and Sun until the overthrow of the shah in 1979. The ICRC officially accepted deviations from the red cross symbol in 1929.[4]

While the number of Red Cross societies rapidly increased, the creation of an American society was slow in coming. Petitioned three times, the United States government refused to sign the Geneva Convention. The initial reasons for not signing were given by Secretary of State William Seward soon after the Convention and were in response to a note from the French government that urged Washington to adhere to this new addition to international humanitarian law. The Convention's rules, Seward stated, already were voluntarily observed by the United States in its treatment of wounded. But, U.S. suspicion of European politics and the disdain for any entangling treaty dealing with war were both in the background. Seward expressed it politely to the French:

> It had always been deemed at least a questionable policy, if not unwise, for the United States to become a party of any instrument to which there are many other parties. Nothing but the most urgent necessity should lead to a departure of this rule. It is believed that the case to which your note refers is not one which would warrant such a course.[5]

Clara Barton, the world-renowned heroine of relief efforts during the American Civil War, happened to be in Geneva resting and visiting friends in 1869. A Red Cross delegation approached Barton, urging her to organize the Red Cross in America and to use her influence to win U.S. government adherence to the Geneva Convention. Until that time, Barton had never heard of the Red Cross, but she became its champion at home, its first national society president, and, through tireless efforts, she persuaded Washington to ratify the Convention in 1882. By 1900, there were Red Cross societies in 37 countries. The number reached 170 by early 1997, and more were expected as the national societies became recognized by the new governments of lands formerly part of the Soviet Union and Yugoslavia.[6]

The Red Cross officials in Geneva also promoted the expansion of the original Geneva Convention by successfully sponsoring further drafting conventions. The first extended protection to victims of warfare at sea (1907), the next to prisoners of war (1929), and finally four Geneva Conventions (1949, with two additional Protocols of 1977) gave the Red Cross expanded authority to provide more effective protection and assistance to the victims of armed conflicts. Today, 188 states are signatories to the 1949 Conventions.

States that accede to the 1949 Conventions legally recognize the ICRC and its right to provide relief during armed conflicts. The Conventions confer a right upon the Red Cross to visit prisoners of war and civilian internees during international armed conflicts. In addition, Article 3, which is common to all four Geneva Conventions, gives the ICRC "a right of initiative" to offer its services in civil wars. The protection of and assistance to civilian victims of war, a new emphasis given in the 1949 Conventions, arose from the ICRC's concern over the large proportion of civilian casualties and civilian persecutions during the Spanish Civil War and World War II. In 1968, ICRC Vice President Jean Pictet raised the issue of internal disturbances and tensions as guerrilla and national-liberation wars swept what was then known as the Third World. This led to Protocol II in 1977, which gave de facto combatant status to guerrillas and gave the ICRC the right to offer its services to a government without the offer's being seen as interference in that state's affairs.[7] Actually, even when a state is not a party to the 1949 Conventions and 1977 Protocols, the Red Cross has made informal requests to offer its services during internal strife. The ICRC believes that international law greatly facilitates relief, but that the absence of law should not prevent assisting the victims of armed conflicts. The ICRC also makes the case that its role in war relief is now firmly embedded in customary international law.

Before one looks at the organization and activities of the entire Red Cross movement, it is worth reflecting on the ICRC's guiding principles because they genuinely unite the entire movement and shape its worldwide activities. First, the Committee is Swiss and only Swiss. Its 15 to 25 members who make policy (see Appendix A for the ICRC's current membership) reflect a state that so values its independence and neutrality that it has so far rejected UN membership. The Red Cross resides in a congenial home, holding as its "Fundamental Principles: humanity, impartiality, neutrality, independence, voluntary service, unity and universality," which are similar principles to those of the Swiss state.[8] The three most political principles have precise meanings. *Impartiality* means relief will be provided to whomever needs it; *neutrality* designates that ICRC workers will not take sides in any armed conflict; and *independence* means that relief operations will be conducted separately from the political control of the warring parties or even from UN peacekeepers. These principles, as later chapters will show, have substantial political effects.

Another principle must be added: discretion. It also has political effects. Pierre Graber, a former president of the Swiss Confederation and chairman of the diplomatic conference that produced the 1977 Protocols, succinctly states why judgmental political pronouncements would damage the Red Cross's neutrality, impartiality, and independence. "Discretion is prerequisite for the ICRC to gain the confidence of all, that confidence which is essential for it to maintain the universal nature of its activities." Similarly, the Swiss Federal Council applies the same "discreet approach to morally reprehensible acts by other governments. The Federal Council has always considered such discretion necessary to uphold the political credibility of this country's neutrality."[9] Such principles make Swiss national interests appear to be no interests at all. Of course, this is an illusion, just as the ICRC's non-political stance is an illusion. Both, also of course, are functional illusions. To be non-political for the Swiss government and the ICRC is to maximize both entities' political impact.

The Swiss state, with its neutrality, impartiality, and independence, actually expresses a national interest that tells other states:

1. Do not look here for a wartime ally; our terrain is not hospitable for decisive war and we will fight to preserve our neutrality.
2. We will do business with anyone.
3. If you want a haven, especially for your money, here it is, with few questions asked.
4. We nurture international law by being the center for law-making conventions using the headquarters of the old League of Nations, now the home of a number of UN agencies.
5. If you are wise and witness our peace and prosperity, then why not do as we do? Please make less war and abide more by international law. The Swiss Confederation actively promotes international law.

The ICRC, to continue the parallel, also pays particular attention to international humanitarian law. Not only does law give the ICRC very special authority in armed conflicts, but the more governments and other warring parties adhere to the law, the easier the relief efforts of the ICRC will be. And, perhaps most importantly, the more international humanitarian law is followed, the greater will be the political effects of international relief efforts towards "civilizing" (and ending) wars.

International law commands states, once they consent to abide by it, to behave in particular ways. The 1949 Conventions updated and expanded the coverage of previous Conventions. They command governments to protect wounded, sick, and shipwrecked members of the armed forces, prisoners of war, and civilians. They also command that governments not force the Red Cross workers in the field to take sides, exclude certain victims from their care, or subject the bearer of the Red Cross emblem to attack. If the Conventions are followed, then the ICRC can realize its principles of neutrality, impartiality, and independence while looking after the victims of conflict. According to Claudio Caratsch, the former vice president of the ICRC,

> a general lack of respect for the provisions of international humanitarian law is now the basic obstacle to the ICRC's work. Where the

> protective emblem is not respected and where the rules of the Geneva Conventions are unknown or simply dismissed, protection can be provided only on the ad hoc basis of special agreements.[10]

The ICRC has relentlessly authored and disseminated international humanitarian law that constrains the combat of warring parties in at first two and now three ways. First, (as we will analyze in some depth later), the ICRC has a record of sponsoring laws that make wars less inhumane. Second, it has sponsored laws that separate state foreign policies from the operations of relief agencies, and therefore these laws allow NGOs to provide humanitarian assistance unimpeded. The ICRC even publicly separates its humanitarian interests from the actions of the Swiss government when Bern becomes more involved in European affairs. If that separation can be maintained, then states cannot manipulate the NGOs who protect and assist the victims of war. States, of course, want to. The recent "humanitarian intervention" (the phrase itself connects relief efforts by states to foreign policies that take sides in war and is thus treated with ambivalence by NGOs) in Somalia, Liberia, and the former Yugoslavia illustrates that the violation of international law by UN peacekeepers, the U.S. military, and other armed forces impeded relief efforts. But we are getting ahead of our story. Before examining the political effects of ICRC operations, including the third way it constrains combat in order to make wars dysfunctional, we need to first see how the entire Red Cross movement is organized and what it does in the field. A description of its organization and operations begins to explain why it has the political effects it has.

ORGANIZATION OF THE RED CROSS MOVEMENT

What is officially called The International Red Cross and Red Crescent Movement is comprised of the ICRC, the National Red Cross and Red Crescent Societies, and the International Federation of Red Cross and Red Crescent Societies. The division of labor amongst the three has evolved and is somewhat confusing. Until the Balkan wars beginning in 1912, the ICRC recognized new Red Cross National Societies; provided some liaison between them; promoted international humanitarian law; and organized and coordinated humanitarian relief to the war wounded, mainly through the National Societies in the field. The Balkan wars proved that National Societies, while they worked well with the medical services of their own country, could not or were not allowed to bring relief to the wounded of the opposing side. They were too national to be neutral or impartial, at least that was the feeling of the warring parties.[11] Thus the ICRC itself took on the administration of protecting and

The Headquarters of the ICRC in Geneva.

Occupying a former hotel on Avenue de la Paix, the headquarters perches on a hill across the street from and at a higher elevation than the United Nations complex. The ICRC appears detached from politics, yet geographically close enough to observe the behavior of the international community. In addition, the headquarters' commanding view seems to suggest its secret wish to influence those below. *(Photo by Nicholas O. Berry)* assisting the victims of conflicts. It would be a neutral intermediary between the warring parties, while still relying on National Societies as partners in relief. The Red Cross won high praise for its massive efforts in World War I.

In 1919, with the establishment of the League of Nations and the expectation that world peace was at hand, the National Societies shifted their attention to the peacetime relief of suffering from national disasters, training in health and first aid, and to the coordination of aid during international natural catastrophes. The League of Nations, it was hoped, would bring a more unified world. This meant that coordination and liaison among the National Societies would need to be strengthened. The National Societies first went to the ICRC, requesting that it be their official liaison body. They insisted, however, that if the ICRC agreed to their request, representatives of the National Societies should join the ruling committee of the ICRC. It was an attempt to internationalize what many National Societies saw as a too-Swiss organization. The ICRC rejected their request, wanting to remain all Swiss, independent, and neutral. The ICRC believed that including representatives of National Societies in its policy-making process would make the ICRC the target of political pressures from the National Societies. The National Societies then created their own liaison body. Under the leadership of Henry P. Davison, president of the War Council of the American Red Cross, the League (now Federation) of Red Cross Societies was established to coordinate international relief. In 1928, a deal was struck, and the rough division of labor that was established then continues. The National Societies would have primary jurisdiction on relief from natural disasters, and the ICRC would be responsible for war relief.[12]

The National Societies, with 128 million members and 275,000 employees, act as auxiliaries to their public authorities. With a combined expenditure of $20 billion, they provide emergency relief with health services, food, and shelter to the victims of floods, earthquakes, and storms, as well as the tracing of these victims. In wartime, they can serve as auxiliaries to the military medical services, can aid prisoners and refugees, and can trace the missing. The Federation organizes and coordinates international disaster response and the establishment of new National Societies. The Federation Secretariat is also in Geneva and has a staff of about 250 from over 30 countries.

The ICRC's main responsibility is caring for the victims of armed conflict, both military and civilian. In wartime, it mobilizes and cooperates with the Federation and National Societies. The exact division of labor is worked out at the International Conferences of Red Cross and Red Crescent, where everybody, including the representatives of states that have ratified the four Geneva Conventions, gets together to review the Movement's past, present, and future organization and activities. Normally, the International Conference is held every four years at various international locations. The last Conference, the 26th, met in Geneva in December 1995. Between Conferences, a Council of Delegates made up of representatives of the Movement's three constituent bodies work on resolutions to guide the Movement and resolutions to be considered at the International Conference. The 1993 meeting of the Council of Delegates, for example, produced and approved a Code of Conduct for NGOs taking part in international relief operations. Chapter 3 looks at this Code in some detail because how the ICRC and NGOs go about their wartime operations determines their political effects. First, let us consider what the ICRC does in the field. Its activities, together with those of the Federation, have three times won the Nobel Peace Prize: in 1917, 1944, and 1963.

OPERATIONAL ACTIVITIES OF THE ICRC

In its 1995 *Annual Report,* the ICRC begins the review of its operational activities with the topic: "Relations with International Organizations." The two subheadings reveal its priorities: l) "implementation of international humanitarian law and support for the ICRC," and 2) "concerted action and preserving the neutrality and impartiality of humanitarian action." Above all, when dealing with international organizations, the ICRC defines its primary role as that of a lobbyist on behalf of international humanitarian law. The ICRC proclaims that it shares the UN aim "to cast out the demons of war."[13] It does so by attending meetings of governmental international organizations, and urging resolutions there that respect the law and reaffirm the ICRC's special role in humanitarian action. In 1995, the ICRC sent delegations to the Heads of State Conference of Non-Aligned Countries, the Organization of African Unity (OAU), the Organization of American States (OAS), the Inter-Parliamentary Union (IPU), the Organization on Security and Cooperation in Europe (OSCE), the Organization of the Islamic Conference (OIC), and, on a continual basis, the General Assembly of the United Nations.[14]

The ICRC successfully lobbied Italy to submit a resolution to the General Assembly granting it observer status. With the co-sponsorship of 132 members and UN Secretary-General Pérez de Cuéllar's endorsement, the resolution passed in October 1990 by consensus. "We are not planning to make speeches at the United Nations," stated ICRC's Claudio Caratsch. "What we want to do is mainly to cooperate in the Commissions and to have access to people attending the General Assembly."[15] In fact, the committee's president, Cornelio Sommaruga, did address the General Assembly in 1994, urging the necessity of "translating into action the obligation to respect and ensure respect for humanitarian law."[16]

Observer status facilitated the sponsorship of an international symposium in Geneva on humanitarian action and peacekeeping operations. The ICRC message there: UN peacekeepers ought to obey international law and respect and allow NGO relief personnel to operate without restraint. Observer status also cements links to UN agencies that the ICRC must deal with in the field—in particular, the new Department of Humanitarian Affairs (DHA), the UN High Commissioner for Refugees (UNHCR), the World Food Program (WFP), UNICEF, and UNESCO. The message: let's cooperate so we can "achieve greater complementarity with certain United Nations programs and agencies," and we will be willing to coordinate humanitarian assistance in the field if UN agencies keep it separate from military action.[17]

ICRC officials in Geneva, at the UN delegation, and in the field get to know everybody who is anybody in international relief, public or private. They practice networking and socializing as a science as well as an art. An outside observer would find the close personal relations between ICRC officials and political leaders to be in sharp contrast with their distant public relationship with the foreign policies these politicians pursue.

This emphasis on law and access to political operatives shows the prime concern of the ICRC: public perceptions and the environment on the ground must be congenial to humanitarian relief. Without them, the ICRC has been forced to reduce, even abandon, relief efforts, as it did for a time in Angola, Sudan, Liberia, Somalia, Bosnia, and Burundi. "Prepare the way" is the sine quo non of effective relief operations.

While the ICRC cooperates with the UN "to achieve greater complementary" with its programs, it continually stresses "the need for an absolute distinction to be drawn between political and military action, on one hand, and humanitarian work, on the other."[18] The main mission, humanitarian action, facilitates all the missions.

The second section in the *Annual Report* deals with "Activities for People Deprived of Their Freedom." The ICRC monitors every situation of armed conflict, international or internal, including internal disturbances or crises liable to cause humanitarian problems. Its personnel visit detainees to determine their conditions and treatment, and their findings are passed on confidentially to the holding authorities. In 1995, the ICRC visited 146,585 detainees in 58 countries.[19] Prisoner release and repatriation were managed in Bosnia-Herzegovina as requested by the parties to the 1995 Dayton Accord. Major programs of prison visitations continued in Rwanda, Liberia, Sri Lanka, Morocco/Western Sahara, Afghanistan, Algeria, Turkey, Chechnya, autonomous Palestinian territories, Israel, Lebanon, and Georgia. In all areas of

The ICRC Museum in Geneva.

These statues represent human helplessness. They stand outside the entrance to the museum as reminders that war can leave people bound, blind, forlorn, and at the mercy of warriors. What better way to free them and anyone like them than to lessen the scourge of war. Besides this message, the museum chronicles the ICRC's over 130-year history of humanitarian efforts in war relief and protection. *(Photo by Nicholas O. Berry)*

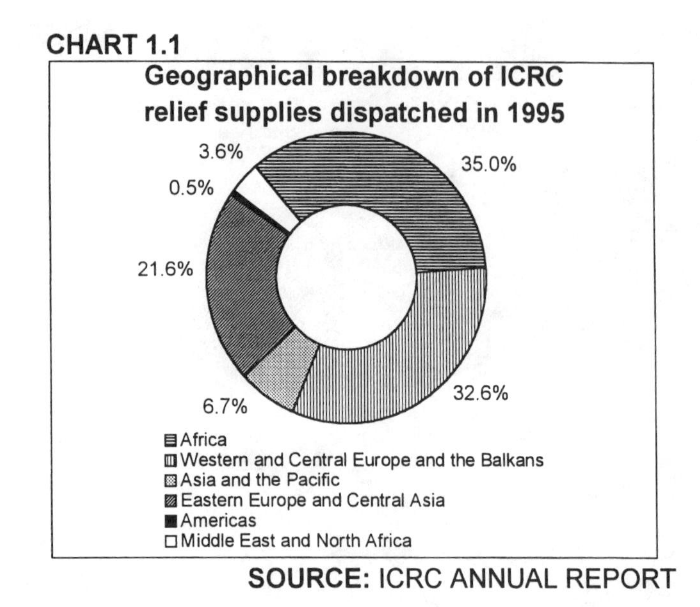

hostilities, the ICRC called on the warring parties to respect international humanitarian law and the immunity of the civilian population.

The Central Tracing Agency of the Operations Department performs the third set of operations. It seeks to unite families and inform relatives of the whereabouts of prisoners, internees, stranded civilians, and refugees. Over 3 million messages were sent in 1995, and over 11,000 families were reunited, both activities taking place largely in the former Yugoslavia and Rwanda.[20] The Agency maintains a vast network of workers specializing in tracing through the National Societies. In Rwanda, 37,000 unaccompanied children were registered in 1994 and, as elsewhere, their names were circulated on flyers and broadcast on radio.

Health is the fourth area of operational activities in meeting the needs of war victims. The Medical Division provides medicines, surgery, orthopedic rehabilitation, water and sanitation, and nutrition training.

The fifth area, relief, involves purchasing, food aid, transport, agronomy, construction, and veterinary medicine. The General Relief Division delivered over 100,000 tons of material in 1995, with 66 percent of it going to operations in Africa, mainly in Rwanda and Angola; and another 33 percent to the Former Yugoslavia, Afghanistan, and the Caucasus.[21] 1995 saw a reduction in the dispatch of relief supplies compared to 1994, during which 206,000 tons of material were provided when the wars in the Former Yugoslavia and Rwanda were at their worst.

CHART 1.2

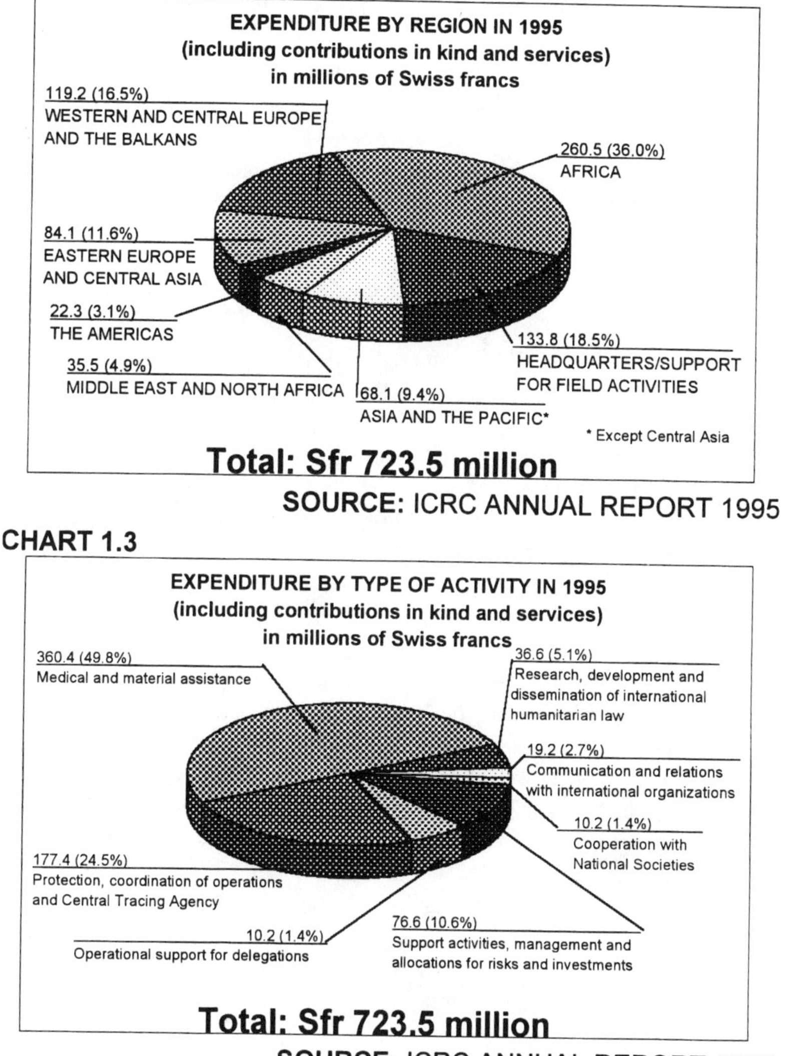
EXPENDITURE BY REGION IN 1995
(including contributions in kind and services)
in millions of Swiss francs
119.2 (16.5%)
WESTERN AND CENTRAL EUROPE
AND THE BALKANS
260.5 (36.0%)
AFRICA
84.1 (11.6%)
EASTERN EUROPE
AND CENTRAL ASIA
22.3 (3.1%)
THE AMERICAS
35.5 (4.9%)
MIDDLE EAST AND NORTH AFRICA
68.1 (9.4%)
ASIA AND THE PACIFIC*
133.8 (18.5%)
HEADQUARTERS/SUPPORT
FOR FIELD ACTIVITIES
* Except Central Asia
Total: Sfr 723.5 million
SOURCE: ICRC ANNUAL REPORT 1995
CHART 1.3
EXPENDITURE BY TYPE OF ACTIVITY IN 1995
(including contributions in kind and services)
in millions of Swiss francs
360.4 (49.8%)
Medical and material assistance
36.6 (5.1%)
Research, development and
dissemination of international
humanitarian law
19.2 (2.7%)
Communication and relations
with international organizations
10.2 (1.4%)
Cooperation with
National Societies
177.4 (24.5%)
Protection, coordination of operations
and Central Tracing Agency
10.2 (1.4%)
Operational support for delegations
76.6 (10.6%)
Support activities, management and
allocations for risks and investments
Total: Sfr 723.5 million
SOURCE: ICRC ANNUAL REPORT 1995

As Chart 1.1 shows, in terms of expenditure, Central and Eastern Europe were the primary recipients of relief supplies in 1995.

The magnitude of ICRC's international operations is indicated by the size of its staff and its budget. On average in 1995, it had 840 delegates working in the field, assisted by 7,000 local employees under contract and 538 National Society personnel seconded to the ICRC. In Geneva, 645 staff worked at headquarters.

Two-thirds of the ICRC's budget—nearly 750 million Swiss francs (approximately $620 million at the 1996 exchange rate)—is contributed by governments. The United States contributes about 20 percent of that, followed by Switzerland, The Netherlands, Sweden, and the United Kingdom. Supranational organizations, National Societies, and gifts provide the other one-third of the annual budget's income.[22]

Charts 1.2 and 1.3 indicate that total expenditures are primarily directed to Africa, then Europe, and used for medical and material assistance, then protection and tracing.

Finally, operation cooperation with National Societies, a subject not listed with the ICRC's operational activities in its 1994 *Annual Report,* shows the need for more cohesion in the entire Movement in light of the growing role of National Societies in internal wars. The organizational chart for the ICRC is presented in Appendix B.

The ICRC has a definite presence in wars and various kinds of internal armed conflicts. It performs some of the same activities as governments that provide relief services to their peoples. That alone indicates there are substantial political effects from what the Red Cross does. Multiply its presence by ten and you have a very rough estimate of all the relief activities (though largely not in war zones) of all the NGOs worldwide. In war zones, the ICRC plays the prominent role, formally and informally collaborating with other NGOs and with UN agencies. Understanding how the ICRC goes about its work will allow us to analyze what those in Geneva prefer to talk about in private—its political nature. We now turn to its new, unspoken mission.

2

The Unspoken Mission: Undermining Civil Wars

War as an institution has evolved over human history. Its conduct has been particularly sensitive to the nature of the existing international system and to changes in technology. The system reveals the stakes and players in war, and technology influences the tools and strategies the players use. A new international system and advanced technologies have dramatically changed the nature of war.

THE NATURE OF MODERN WAR

Wars in the post–Cold War world are overwhelmingly internal, not international.[1] The explanation for this intriguing phenomenon hinges on the foreign policies of the major powers, which are shaped by the international system and the skyrocketing technological costs and effects of international war compared to internal war. International wars are now largely dysfunctional. Internal wars—so far—are not.

In contrast, wars during the bipolar Cold War system were with few exceptions, international. The key factor was the backing of regional or domestic warring parties by the two superpowers. The wars between states—the Korean War; the 1956, 1967, 1973, and 1982-84 Middle East wars; the Vietnam War; the 1965 and 1971 Indian-Pakistani wars; the 1978 Viet-Cambodian War; the 1979 Sino-Vietnamese War; the Falklands War; and the Somali-Ethiopian War—found the superpowers on different sides.[2] They supported the wars, providing arms, aid, advisors, and—for America in Korea and Vietnam—combat troops. Policies of deterrence, backed by nuclear weapons, limited the escalation of these wars. When wars were fought within states (or colonies)—in the Philippines, China, Greece, French Indochina, Guatemala, Algeria, Hungary, Congo, Laos, Cambodia, El Salvador, Nicaragua, Angola, Afghanistan, and elsewhere—the superpowers were engaged similarly in backing either the government or the insurgents.[3] During the Cold War, the balance of power between the United States and the USSR became the primary determinant of which side would win the struggle. "Internal" wars that affected the balance of power were too important to allow them to remain exclusively internal. The locals were not permitted to fight it out by themselves. Top ICRC officials look back on these internal struggles and refer to them as "international civil wars." Superpower intervention enlarged them and made them a part of the broader international system.

Now, both international and internal wars have a new pattern.

If a regional war between states becomes likely, the major powers can be expected to cooperate and intervene to deter the war or to settle the war if it breaks out. The current international system is highly integrated, which means that the major powers take into consideration the entire system when shaping their foreign policies. In particular, it is now widely accepted that the health of national economies (and the fortunes of the party in power) depend upon the health of the international market and currencies. All major powers support the World Trade Organization (WTO) or, as with the Chinese and Russians, want to join. No major power plots to overthrow what is commonly called the interdependent multipolar system. No major power is prowar. In fact, the major powers will intervene to stop wars between states in order to protect markets, stop genocide and mass murder, avoid floods of refugees, manage nuclear non-proliferation, and prevent the war from spreading. Under the aegis of the UN Security Council, the major powers are able to employ arms and economic embargoes, propaganda, and armed peacekeepers to contain or resolve any regional, international war. The interests of the major powers would weigh heavily in their settlement. This happened in Desert Storm after the brief war between Iraq and Kuwait. No international wars of any magnitude in any region can now be isolated. The major powers, if necessary, are able to shape the cost-benefit calculations of potential belligerents by increasing economic costs and decreasing the likelihood of victory. This discourages the militancy of leaders of small states. War then fades as a strategic option for settling disputes with neighbors. The future of regional, international wars is bleak.

The current multipolar international system also does not favor wars between the major powers. Multipolarity, as a system, indicates that no security threat exists for the big states. The sides for a possible war have not formed. And if one big state gets in a warlike mode, such as a Russia might under a neofascist leader, the other major states will coalesce to deter that possible aggressor. More major states now have deterrence-enhancing nuclear weapons.

International wars have also become extremely expensive. One of the lessons that Desert Storm taught is that advanced technology—in weaponry, surveillance, electronic warfare, and communication—gives the superior side a disproportionate advantage. This favors the major, status quo powers, especially the United States. But the costs of maintaining an advanced military establishment are high. A heavy bomber in World War II cost about $200,000; the price tag on a B-2 today is over $1 billion. First-class battle tanks cost over $1 million. Modern warfare consumes weaponry at a prodigious rate as more effective defensive weapons engage more effective offensive weapons. The last two Arab-Israeli wars ran out of steam when paying the costs of weapons could not be sustained. Even the large oil revenues of Iran and Iraq eventually became insufficient to fund their 1980-88 war. The Coalition forces in Desert Storm, in effect, became mercenaries when they were compelled to seek financial support from Japan, Germany, and other wealthy states. It is a very new question in human history. Who can afford to conduct modern, international war?

Finally, democracies, for the first time in history, comprise more than half of the world's governments. Democracies do not make war on each other for a variety of reasons: It is hard to demonize a democratic enemy government that is an expression of its people's will; democracy is a process of reconciling policy interests through compromise, and democratic leaders expect to use the same process among each other in resolving international disputes; and democracies can only generate grand war goals against dictatorships, because to use war to force the submission of other democracies is to rob their people of policy influence. Never before has war been less attractive as a method of settling disputes between states. What is called "The Long Peace" will get longer.

ICRC officials, in my experience, analyze international politics as well as anyone. They are not overly concerned with the future burden of international war upon their humanitarian work. In fact, international wars were not even mentioned (except in reminisces) at the 1995 International Conference of the Red Cross and Red Crescent. The ICRC is, however, highly concerned with the explosion of very nasty internal wars. That is its overwhelming focus.

The Cold War, while it internationalized wars, repressed many internal disputes over religious and ethnic rights. An external threat, such as that from the Soviet Union towards Yugoslavia, sublimated domestic differences. Unity was demanded by governments for good reason. Splits would invite superpower intervention. In the case of the allies or clients of superpowers, unsettled domestic animosities—such as tribal differences in Somalia and Liberia—were papered over with the help of the superpower patron.

The end of the Cold War ended most foreign threats. One of the motives for domestic unity disappeared. The end of the Cold War also ended major-power concern for political correctness among allies and clients. Governments no longer had to be propped up to preserve unity among the Free World or the Socialist Camp. Governments were on their own. The lid was off. Some governments, such as Argentina, Nicaragua, South Africa, and Czechoslovakia (by splitting), handled their internal problems and resolved differences. Others did not. Internal differences exploded into war. Even the wars in the former Yugoslavia, in

Bosnia and Croatia, can be described as internal despite the infiltration of national personnel and war material into both conflicts. There and elsewhere—in the Sudan, Somalia, Sri Lanka, Cambodia, Rwanda, Azerbaijan, Georgia, Liberia, Chechnya, Indonesia/East Timor, Algeria, and a dozen other places—extreme ethnic, religious, and national differences produced civil wars and revolutions.

These internal wars, the ICRC believes, arise from identity crises within states where governments lack the authority to create and preserve unity among their peoples—a unity once enforced during the Cold War. The "disintegration of state structures" is a common feature.[4]

Compared to international wars, these internal wars are inexpensive to conduct. The air forces and heavy weapons that are necessary to maintain front lines in international wars are usually not needed in civil wars and revolutions. While front lines are not irrelevant in internal wars, as those in the former Yugoslavia attest, they are not simply territorial wars. Instead, internal wars intimately involve civilians. For combat among civilians, cheap and abundant AK-47 assault rifles are more usable than bombers. These cheap wars, so far, remain options to settle domestic disputes.

Internal wars can even be labelled civilian wars. How civilians are mobilized and the degree to which they support their fighters in the field will largely determine the outcome of these struggles. "The fish are the fighters and the people are the sea," said Mao. This makes civilians prime targets of attack. Evaporate the enemy's sea and its fish die. There are even new names for this combat focus on civilians: religious purification and ethnic cleansing. What is new, although not unprecedented considering the Thirty Years' War, is the greater number of civilian deaths over those of the armed forces. The ICRC estimates that civilians make up 90 percent of those directly affected by fighting, including casualties.[5] Failure to support one's group as well as loyalty to another group can bring death. Civil wars are vicious by their very nature. There is no such thing as a civil, civil war.

Another key indicator of the vital role of civilians in internal wars is the explosion in the number of civilian refugees. According to the UN High Commissioner for Refugees (UNHCR), the first four years of the wars in Bosnia and Croatia produced an estimated 2.7 million refugees.[6] When Croatian forces captured its Serbian-populated Krajina region in August 1995, 220,000 Serb civilians fled. No one knows exactly, but ICRC estimates put the number of worldwide refugees in 1995 at over 23 million,[7] and the UNHCR put it at more than 27 million.[8] In effect, creating refugees is the draining of the sea. It is testimony to the importance of civilians in internal wars.

Finally, there are often more than two warring parties in today's wars, as is true in Sudan, Afghanistan, Liberia, Lebanon, and Somalia. Chaos is a fact of

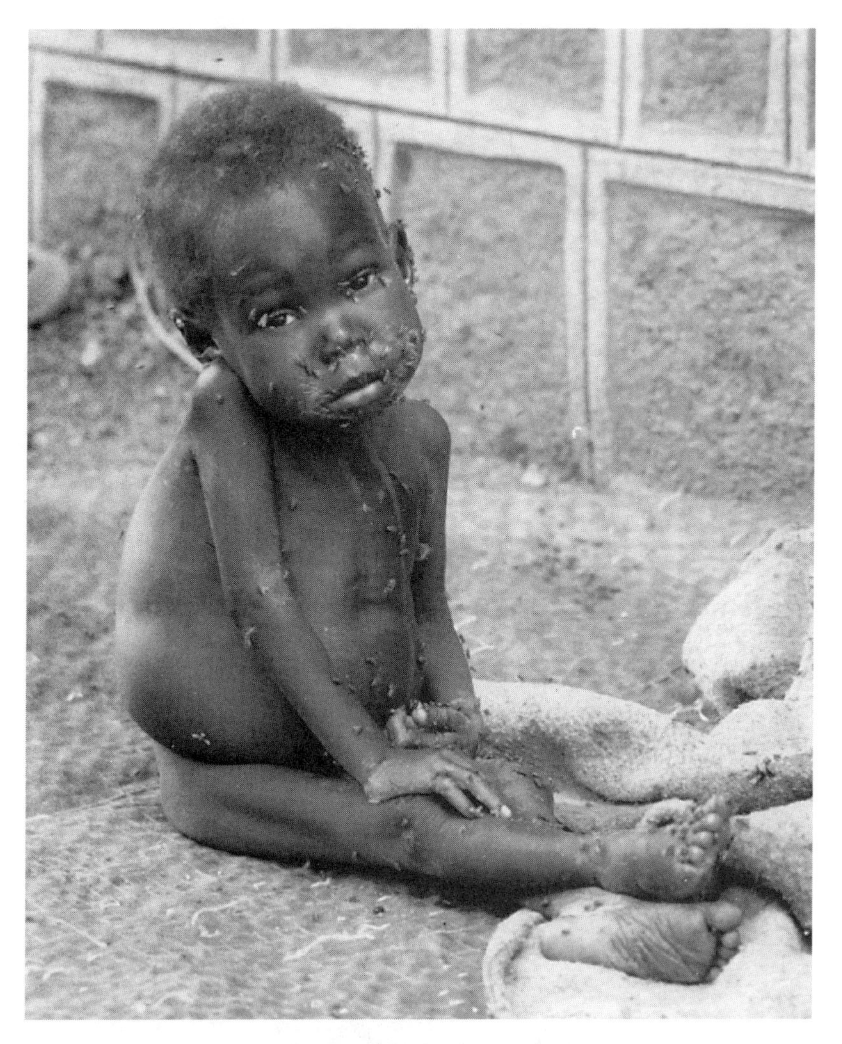

A Young Victim in Sudan.

A photograph like this cannot help but raise questions in the minds of viewers. What happened to cause this child's abandonment? Who will care for this little person? Is he (or she?) alive today, and, if so, what physical and mental scars have been borne? The ICRC makes photographs showing the consequences of war freely available to the media. These sad portrayals serve to shape attitudes in support of the work of the ICRC and against the nasty business of war. *(Photo courtesy of ICRC, by Thierry Gassman)* life and a cause of death, made worse by warlords who play on ancient hatreds and employ violence-begetting-violence strategies to keep themselves needed in power.

THE PROBLEMS NGOs FACE IN DELIVERING RELIEF IN INTERNAL WARS

Humanitarian NGOs are in the midst of the civilians, and they have found their personnel as vulnerable to assaults as those they serve. Since 1985, the ICRC has had 18 expatriates and over 40 local employees killed and 147 "disappeared."[9] Red Cross and Red Crescent workers have recently been killed in the line of duty in Burundi, Chechnya, Colombia, Rwanda, Turkey, and Zaire. Some of the worst attacks on aid workers occurred during the genocide in Rwanda in 1994. The ICRC's delegation and hospital in the capital Kigali repeatedly came under fire, and both staff and patients were killed. Red Cross ambulances were stopped and the patients inside murdered.[10] Daniel Augstburger, an ICRC delegate with long service in Africa, was driving one of these ambulances that the Hutus turned into a meat wagon for Tutsis. He expressed to me his profound sorrow and frustration that he could do little to save the wounded in his charge. In neighboring Burundi, three ICRC delegates were fired upon and killed in their vehicle on June 4, 1996, after delivering water, medicines, and supplies to an isolated area hit hard by ethnic fighting.

The highest death toll in ICRC history occurred in war-ravaged Chechnya on December 17, 1996, when six delegates were murdered in their beds at an ICRC hospital outside of the capital, Grozny. (Chapter 8 will analyze this tragedy in detail.)

In Somalia, Sudan, Bosnia, and Angola, various warring parties have demanded and received "protection" money from NGOs and payments for allowing the distribution of relief supplies, even to their own people. The warring parties routinely block the distribution of food and access to refugees and prisoners. In Bosnia, for example, aid convoys to the protected areas of Bihać and Gorazde were often blocked for weeks.[11] The same thing happened in Sri Lanka, where the government blocked all relief agencies from delivering aid to the rebel-held Jaffna Peninsula during its fall 1995 offensive against the Liberation Tigers of Tamil Eelam.[12] Soon after the offensive and after many civilian casualties, the ICRC was allowed to continue its relief efforts. The struggle for power in Liberia became so horrendous that the ICRC and other aid organizations were forced to suspend operations outside of Monrovia. Their workers were "harassed and provoked by armed individuals and terrorized by displays of extreme violence, including killings, cannibalism and complete disrespect for the mortal remains of victims."[13]

UN and other peacekeepers have also interfered with relief efforts in the field, usually for providing humanitarian assistance to the "wrong" side. Blue helmet peacekeepers actually fired upon hospitals and aid organizations in Somalia. In Liberia, the West African peacekeepers (from ECOMOG, the Monitoring Group of the Economic Community of West African States), twice destroyed aid convoys because they were delaying the "final victory" of the peacekeepers.[14]

All of these activities are against international humanitarian law and greatly hinder NGO relief work.

The ICRC sees some of the vast quantity of weapons produced during the Cold War finding their way to the warring parties. "Even bandits are well supplied," laments the author of a working paper used at the 26th International Conference of the Red Cross and Red Crescent. The report continues:

> Civil and inter-ethnic wars are devastating vast tracts of land, reducing them to chaos and anarchy fraught with extreme insecurity. With time, the social fabric is destroyed, any form of authority, except that of the gun, completely disappears, and fundamental values are denied. In many conflicts the warring parties' political objectives are being replaced by hatred, banditry and the arbitrary wielding of power. From the outset, many confrontations seek first and foremost to destroy the other side, for reasons of racial, nationalistic or religious extremism, or even for economic reasons.[15]

The genocide in Rwanda, for example, was perpetrated in many instances by killers not yet in their teens. Rebel forces are untrained, frequently undisciplined, often without a reliable chain of command, and hard to identify. Government forces tend to behave similarly, saying that they are fighting fire with fire. The rules of war found in the four Geneva Conventions of 1949 and the two Additional Protocols of 1977 are not accurate predictors of how fighters will behave. ICRC officials frequently call today's internal wars "unstructured conflicts."

Yves Sandoz, ICRC director of Principles, Law and Relations with the Movement, expresses anger at the barbarity of current wars and concludes that it is totally unacceptable.

> Every day children are blasted by mines, forced into combat or driven to prostitution as a means of survival; every day, women are raped by soldiers, separated from their kith and kin or left alone to fend for themselves and their infants; every day, prisoners are tortured or allowed to waste away in their cells, sometimes long after the end of the conflict that led to their incarceration; every day, the

Child Soldiers in Afghanistan.
According to the laws of war, the legal age for warriors in combat is fifteen. But many children—who are generally more easily influenced by authority than adults—are fighting. The death of children always and inevitably injects more hatred into the conflict. The ICRC wants the legal minimum age to be raised to eighteen, giving these children the chance to understand the beauty of life and to question those who would order its destruction.
(Photo courtesy of ICRC, by Didier Bregnard)

> inhabitants of entire areas have no choice but to flee from their homes. All this suffering is the result of hatred, stupidity, arrogance and probably also despair.[16]

Not all NGOs responded in the same way to these tribulations. Médecins sans Frontières (MSF, or in English, Doctors without Borders) has favored maximum publicity against the violators of international law, including the conduct of operations on the rebel side, if it is the virtuous party, without the consent of the local government. The ICRC, while it also extensively employs publicity—including sponsoring a book, *Crime Without Punishment,* cataloging the brutality in the former Yugoslavia—has tried to maintain its principles of impartiality, neutrality, and independence.[17] These principles maximize its access to internal wars, and thus maximize its relief efforts and chances for success in its unspoken mission: the undermining of these horrible wars.

THE BIRTH OF THE UNSPOKEN MISSION

The idea of the unspoken mission has been in the minds of Red Cross leaders almost from the beginning of the Movement. But it was a distant dream back then. Founder Henry Dunant, explaining to a friend why he had written *A Memory of Solferino,* wrote that he wanted to instil in people world-wide "a religious horror of war and thereby converting them into friends of peace."[18] Dunant saw the shaping of antiwar public opinion and the establishing of mandatory arbitration of international disputes as the two principal ways to struggle against war. His cofounder, Gustave Moynier, came even closer to defining today's unspoken mission. In a newspaper article written in 1892, Monyier answered to the charges that promoting peace—not caring for the victims of war—should be the primary concern of the Red Cross. He began by noting that warfare "is likely to hold sway in the world for a long time to come." Thus, war relief would remain the most urgent task. But, he wrote, "[even today] the collective conscience of humankind . . . has reached the conclusion that war must be considered a morbid disorder that must be eliminated." The ICRC hadn't given up the fight against war, Monyier continued, but fought against it in a different way than the preaching of peace societies. "We have tempered combat by imposing a few not too troublesome restrictions" through the work of the Red Cross in the field and through efforts to establish international treaties that restrict warfare. Together, Monyier believed, they would create a situation that "will evolve towards [the] gradual elimination of the use of violent means for the settlement of international conflicts."[19]

Recently these ideas have been given new life. The brutality of post–Cold War wars made them intolerable to ICRC leaders. If there were no policy choices but to provide relief while stoically bearing the horrors of war, then the ICRC would have done just that. But ICRC officials searched for a broad policy alternative and found one. Their studies in the early 1990s, a time during which the brutality of internal wars came sharply into focus, led them to the conclusion that more intense humanitarian assistance and political intervention could protect civilians and make these wars dysfunctional. This was one theme that emerged from the Red Cross–sponsored 1993 International Conference for the Protection of War Victims.

It is worth mentioning that another study group, meeting at the same time, came to the same conclusion. Organized in New York by The Council on Foreign Relations, the group chose "Collective Involvement in Internal Conflicts" as its subject. Lori Fisler Damrosch, codirector of the group, described in her introduction to the group's report an important finding:

> The title of our volume—*Enforcing Restraint: Collective Intervention in Internal Conflicts*—reflects the complexities of our subject. Above all, it suggests the emergent capabilities of the international community to restrain those who are bent on destruction and oppression within their own societies. "Intervention"—a controversial term with multiple connotations—here refers to affirmative international efforts to separate combatants and enforce peace.[20]

In short, a close look at internal wars revealed how fragile they were. Unlike World War II, Korea, Vietnam, or even Afghanistan, where the presence and relief activities of the ICRC workers were overwhelmed by the scope of these wars and thus had little effect on their outcomes, today's wars are vulnerable to intervention. They are small wars, unsupported by superpower patrons. Inject members of the international community into these wars as non-partisan third parties, and these rag-tag wars can be overwhelmed and made dysfunctional. They can be sabotaged. The third parties would have, in most cases, enough capabilities to upset the power relations between the warring parties and prevent victory by one side, to mute the horrors of combat on civilians, and to create the basis for a diplomatic settlement.

Early indications that this analysis was correct came from the settlements of a handful of wars in the post–Cold War period. Only one war resulted in a traditional victory. Except for the Eritrean war of independence (and even there the costs were horrendous), not one produced even a rough attainment of goals by any side. A measure of dissatisfaction with wars' outcomes affected all the warring parties. The wars in El Salvador, Nicaragua, the Persian Gulf, Mozambique, Angola, and now in the Balkans and Guatemala produced highly compromised endings. Third-party interference upset the usual process of war and its ability to resolve conflicts with victory for one side. War as a strategy appeared to be weakening.

In 1995, ICRC officials began to be less cryptic about their unspoken mission—but still did not disclose it. ICRC President Cornelio Sommaruga, in his keynote address to the 26th International Conference of the Movement, reflected on the history of the Red Cross role in war. "World War II was a setback for the Red Cross," he said. "We did not have the power to stop its inhumanity." Even with the A-bomb and the creation of the UN "to stop the holocaust of war," the assumption was that "war would not go away." "Now," Sommaruga declared, "there is optimism." He proclaimed that "humanitarian action must be in parallel with political action." Together, "a new contract of humanity" would find international humanitarian law applied to all armed conflicts, and, presumably, the end to "barbarity we thought belonged in the distant past."[21] It is not a direct assault on war, but close to it.

Even more explicit, but with a disclaimer that his views "do not represent the views of the ICRC," Kim Gordon-Bates, an ICRC public information official, made no secret of his repugnance for war and of his ideas what should be done to remedy its criminal nature. Writing in *Crosslines,* Gordon-Bates presents a policy to remedy the dilemma

> if the various and precise articles of the Geneva Conventions *et al* are not able to influence or contain the conduct of, say, a war of extermination and if the states continue to confirm their impotency. Surely then, we must grasp the spirit of humanitarian enterprise that led to their creation and recycle it differently. This means working to prevent, and not simply regulate, conflicts—particularly those liable to burn beyond the pale of reasonable authority.

He goes further:

> Torture, slavery, collective guilt, public executions . . . became unacceptable and today, in the latter half of the twentieth century, are no longer regarded as legitimate. When exposed, they are duly condemned. We therefore need to promote a desire to further widen the scope of what human morality should consider "unacceptable" or "criminal."[22]

The brochure that presented the main subjects to be addressed by Commission I at the International Conference began with the basic problem involving "War Victims and Respect for International Humanitarian Law." The world, it stated,

> has been confronted with inexplicable brutality committed in conflicts in nearly every corner of the earth. While these horrors are not necessarily new in the annals of human cruelty, they are brought to us daily on our television screens. They concern us. They reflect the barbarity of our times, and *we bear a collective responsibility to put a stop to them.* (My emphasis.)[23]

Speaking on background, a veteran high official told me that the ICRC "cannot use [only] traditional tools," namely international humanitarian law, to put an end to what he called the brutal "wars of identity" that are "expressions of the failure of the state." Interfering with these wars, admittedly "a bigger role," is not an "official policy, just adapting" when "systems collapse." "Yes, it [humanitarian action] may prolong wars, but it also saves people; and, yes, in the long range, it hampers wars." The ICRC cannot "go public—to a full extent," since doing so would damage humanitarian action and the interests of the victims.[24] This official, and many others interviewed, not only did not deny that a new mission existed, they did not express any disapproval of this unspoken mission. All clearly recognized, however, the sensitive issues raised by pursuing both missions.

Writing in *Crosslines* (put out at the time of the International Conference), Yves Sandoz made an unusual leap, connecting principles and law to putting an end to war.

> The second aim of the Conference is to bring about the adoption of practical measures to increase compliance with humanitarian principles and rules during armed conflicts, and to strengthen and facilitate humanitarian action at all times. Of course, one cannot expect spectacular change at once; *conflicts are not going to vanish overnight.* But urgent needs should not blind us *to the importance of planning for the long term.* (My emphasis.)[25]

When ICRC officials say, and they all say it, that the humanitarian mission comes first, they give it priority because it is their traditional, official, and most pressing mission. The unspoken mission is unofficial, dependent on performing the main mission, and, as Sandoz writes, long-term.

Daniel Augstburger, now an ICRC spokesman at its UN delegation in New York, perhaps tiring of my questions designed to get him to go on record admitting the existence of the second mission, finally conceded with more than a hint. "Do we interfere with the way you [sic] make war—of course." Remaining parsimonious in his words on the subject, he answered my question: What is the future of war? with a terse: "Not much."[26]

The ICRC's strategy to undermine war includes an ingenious method to keep humanitarian action unsullied by politics. On the surface, the ICRC can maintain its principles of neutrality, impartiality, and independence while still supporting the assault on war. The key is to keep the undermining of war on two tracks that appear to be separate, and, in guiding operations, actually are. The best description of the strategy is found in a cover letter and transcript of a talk addressed "to the attention of National Societies" dated November 29, 1994. Yves Sandoz, who appears in ICRC literature as a major policy planner in the Movement, addresses the National Societies, some of whom "expressed interest in ongoing reflections on the role of the Movement in conflict prevention in its broad sense." In his cover letter, he writes:

> I believe that the message the Movement should convey is as follows:
> Our efforts to promote and implement international humanitarian law are not incompatible with those to promote peace and human rights, but on the contrary reinforce them; By working to help the most vulnerable groups, National Societies can play a role, however modest, in the overall efforts that must be made to eliminate the root causes of armed conflict.

Sandoz begins his talk, given at a San Remo (Switzerland) round table and now passed on to the National Societies, by stating a question and then answering it.

> Fight against armed conflict, or limit its horrors? The question which lies at the heart of our discussion today has been asked by the ICRC ever since it was founded: Should the ICRC endeavor to limit the horrors of armed conflict, or fight against it directly?
>
> The answer was provided by Henry Dunant in his well-known book, *A Memory of Solferino,* written in 1862: "If war is inevitable, it should be waged with as little barbarity as possible." In fact, on the basis of Dunant's horrendous experience at Solferino, the ICRC decided to draft a "Convention for the Amelioration of the Condition of the Wounded in Armies in the Field", which was adopted by the signatory States on 22 August 1864.
>
> In setting itself this limited objective, the ICRC was motivated by realism rather than by indifference to the phenomenon of war. As one of its founding members, Louis Appia, wrote the year following the adoption of the 1864 Convention: "Let us loudly proclaim our grief, our pain at being unable to do more, let us protest against the great collective iniquity that goes by the name of war. . . ."
>
> The ICRC has been closely linked to international humanitarian law throughout its history; nonetheless, well-intentioned voices are constantly raised, asking if it might not be better to fight against war itself rather than strive to limit its horrors.

> *There is no dilemma here, none whatsoever.* Nowadays we do not ask whether it is best to promote research on cancer and AIDS instead of improving the care of sufferers of these diseases: both actions are essential. By the same token, *international humanitarian law does not contradict peace efforts but complements them.* It is still far too often necessary to make this clear. (My emphasis.)[27]

Sandoz makes it clear that "conflict prevention and the restoration of peace naturally continue to be the main task of the United Nations." It follows that if the UN and its peacekeepers can be induced to intervene in internal wars and follow rules that would lead to a cease-fire and a settlement, then the ICRC does not have to be overtly political. No dilemma. On a parallel track that "does not contradict peace efforts," humanitarian actions can "reinforce" these peace efforts. It is essential that the two tracks not be publicly linked. Separately, both can head in the same direction.

ICRC officials, therefore, have developed a strategic plan for each track. One plan focuses on international organizations (IGOs)—primarily the UN—and their members. ICRC lobbying efforts towards IGOs are the subject of chapter 4. The second strategic plan targets other NGOs and the National Societies. It seeks to guide their behavior in harmony with the activities of the ICRC. This plan is designed to preserve vital access to war zones, maximize the provision of humanitarian assistance, and at the same time interfere with, mute, and end any internal war. The centerpiece of this strategy is the adoption and promotion of a Code of Conduct for the National Societies and all NGOs involved in humanitarian relief. The next chapter presents an analysis of the Code of Conduct, explaining first how it serves the delivery of humanitarian assistance and then how it serves to undermine wars.

If war as an institution fails to resolve conflicts to the satisfaction of at least one warring party, then it is on the road to obsolescence. What better way to protect the victims of war then to prevent war and its creation of victims?

3

Guiding the International Community

The ICRC traditionally has encouraged other NGOs to follow its Fundamental Principles of humanity, impartiality, neutrality, independence, voluntary service, unity, and universality. With decades of experience, the ICRC knows that following these principles allows its workers to function in the field while minimizing the hazards of their very dangerous duties. The Principles facilitate humanitarian relief. The adoption of another mission prompted the ICRC to reiterate the principles as a set of rules—in order to guide *all the players* in the international community that might intervene in wars. These rules were needed to facilitate lessening the brutality of internal wars and eventually to make them dysfunctional and settled. Fulfilling both missions required a code of conduct.

THE CODE OF CONDUCT

In October 1993, the Movement's Council of Delegates convened in Birmingham, England to adopt a Code of Conduct for NGOs to follow in their humanitarian action. A Code with ten principles was approved. In addition, the Council of Delegates accepted three Annexes that included recommendations: 1) for governments in disaster-affected countries; 2) for governments donating to disaster relief; and 3) for intergovernmental organizations (IGOs) involved in peacekeeping or relief.[1] George Weber, secretary general of the International Federation of Red Cross and Red Crescent Societies, chaired the deliberations. The ICRC wanted the Code of Conduct and the Annexes to be a guideline for the entire Movement. ICRC officials believe that National Societies at times unfortunately tend to support their national government's foreign policy, which, in the words of one official, "puts sticks in our way." Coordinating Movement unity in war relief, the official task of the ICRC, has always been a persistent problem.

The Code of Conduct was drafted by the Steering Committee for Humanitarian Response, which was made up of seven NGOs in cooperation with the ICRC. The NGOs were Caritas Internationalis; Catholic Relief Services; the International Federation of Red Cross and Red Crescent Societies; the International Save the Children Alliance; the Lutheran World Federation;

Oxfam; and the World Council of Churches. By early 1997, a total of 104 NGOs had agreed to follow the Code.[2] More sponsors are being sought.

The Code is designed "to maintain the high standards of independence, effectiveness and impact to which disaster-response NGOs and the International Red Cross and Red Crescent Movement aspires."[3] "Independence, effectiveness and impact" describe the three strategic purposes of the Code. Independence provides NGO access to wars; effectiveness maximizes aid to their victims; and impact seeks to stop the horror of wars. Both the main and unspoken missions are served by the Code of Conduct.

Independence (or at minimum the appearance of independence) from the political work of the UN or the warring parties is necessary to gain access to the victims of war. The International Conference working paper for Commission II on "Humanitarian Values and Response to Crisis" made clear this purpose of the Code.

> [I]t is important to make a distinction between the political responsibilities of States as well as of the United Nations and regional organizations, and the responsibilities attached to humanitarian activities conducted by neutral and impartial humanitarian agencies. There are two quite separate functions involved: one is that of the police and judiciary, stemming from the duty to see justice done, to ensure respect for the law and to punish violations; the other is that of the aid worker, whose sole concern is to protect and assist each and every victim in the name of humanity. The need to reinforce this distinction and ensure that humanitarian agencies respect the basic principles of humanitarian work as well as having access to those in need, was well highlighted at the 1993 meeting of the Council of Delegates of the International Red Cross and Red Crescent Movement.[4]

The more that NGOs are connected to political activities, the more they will be "held responsible for the acts of others and for the results of their politics and strategies."[5] This is to be avoided. A veteran ICRC official explained to me that "a division of work" between NGO aid agencies and UN peacekeepers was necessary to keep the NGOs non-political, thus ensuring NGO access to all sides in the war and the safety of humanitarian workers. As with international humanitarian law, the adoption of rules for NGOs tends to depoliticize their actions. Following preset rules of independence, neutrality, and impartiality is different from following policies based on political decisionmaking. Rules can obviate political choice.

The same veteran official also said that the *effectiveness* of humanitarian assistance depended upon our efforts "to get other actors to work with us.

Many NGOs are newcomers, and are not always neutral or impartial—and they have got good government connections. The warring parties will reject the ICRC offer [of humanitarian assistance] if these NGOs take sides."[6] The ICRC and other neutral NGOs will be shut out by the warring parties because they will have the option of letting in only the NGOs that support their cause and keep out those that would also aid their enemies. Relief work will then suffer. The proliferation of novice (some experienced field workers use the word "nutty") NGOs has been spectacular. For instance, in Kigali, Rwanda's capital, 120 NGOs registered in 1994.[7] This has inevitably increased unprofessionalism and hampered relief.

> The immediacy of disaster relief can often lead NGOs to unwittingly put pressure on themselves, which leads to short-sighted and inappropriate work: programmes which rely on foreign imports or expertise, projects which pay little attention to local custom and culture, and activities which accept the easy and high media-profile tasks of relief but leave for others the less appealing and difficult ones of disaster preparedness and long-term rehabilitation.
>
> All NGOs, big and small, are susceptible to these internal and external pressures. As they are required to do more and as the incidence of complex disasters involving natural, economic and often military factors increases, the need for some sort of basic professional code becomes more and more imperative.[8]

The ICRC, as you can guess, does not name names of those NGOs whom it finds so unprofessional. Instead, its leaders speak of a need for "a new discipline among aid agencies."[9]

While not declared, the *impact* the ICRC desires from humanitarian relief includes more than relief. There is no doubt that by following the Code, NGOs will create a climate more conducive to relief operations. What is nowhere mentioned are the effects on the conduct of war that will come about if the Code and the recommended guidelines to other participants are followed. First, we will review the Code and how it will facilitate humanitarian relief. Then we will suggest what it does not say: that following the principles will make war far more difficult to conduct. The Code for NGOs has ten principles.

1. The humanitarian imperative comes first. This principle states that relief should be enjoyed by all in need, "hence the need for unimpeded access to affected populations." Furthermore, "aid is not a partisan or political action and should not be viewed as such."

Access and the lack of interference in relief operations certainly facilitates aiding the victims of war. The connection between provider and recipient is then direct. The wishes of the warring parties or peacekeepers are irrelevant. It is also

A Family in the Former Yugoslavia Is Reunited.

Wars such as that in the former Yugoslavia break up families and cause incomparable suffering. But war weariness can also become a war retardant, eating away at the vitals of the energy, morale, and hope so necessary for the conduct of war. The ICRC reunites families and, while providing a partial remedy to the victims of war, creates the hope of a better life in peace. *(Photo courtesy of ICRC, by Ana Feric)* important that humanitarian agencies be perceived as not taking sides. Giving aid to all sides is designed to paint that aid as impartial and non-political.

But consider this. If NGOs have an imperative to operate in war zones, to be unimpeded in serving the needs of people, are they not reducing the ability of armed forces to punish the enemy, including civilians who sustain the enemy's forces? Francis Deng, a former Sudanese diplomat, describes the basic problem in unimpeded access: "When a man has placed his own life on the line and is prepared to kill or die for a cause, it is difficult for him to be overly concerned about the humanitarian needs of those who have remained behind enemy lines, especially if that would compromise the cause for which he has chosen to make the ultimate sacrifice."[10] How can breaking the enemy's will to resist, the Clausewitzian imperative, be accomplished if there are NGO workers on the scene who are free to protect, feed, clothe, and nurse prisoners, refugees, and civilians? Proclaiming that aid is not partisan or political does not make it so. The hundreds of NGO relief workers who have paid the ultimate price for their humanitarian calling, while perhaps a risk facing everyone in war zones, testifies that they are sometimes perceived as political.

It must also be said that this "humanitarian-imperative" principle justifies and legitimizes the dangerous role of aid workers in wars—to the warring parties, to any political/military third parties, and to the aid workers themselves. Just as a warrior needs a political cause to "put his life on the line," so too does a humanitarian worker need a cause. Access to war zones may be ICRC's "top priority," but having relief workers willing to be there is also a priority of some importance.

2. *Aid is given regardless of the race, creed or nationality of the recipients and without adverse distinction of any kind.* "Aid priorities are calculated on the basis of need alone." This principle in the Code says that "life is as precious in one part of a country as another." The "degree of suffering" alone will determine the provision of aid.

Clearly, to be effective humanitarian action must be able to go where human suffering is the greatest. This principle, on the face of it, removes any overt political taint to humanitarian action. Need is need.

But consider this. This principle allows relief to go to a warring party's enemy, and the more that enemy is punished, the more relief its people get! This is reminiscent of the American bombing campaign of North Vietnam. Reportedly, for every one dollar of destruction the bombs produced, Hanoi's Chinese and Soviet allies contributed two dollars' worth of economic and military assistance. The bombing was counterproductive. The same is true for combat against the enemy in internal wars under this principle. The more enemy forces and their civilian backers are punished, the more humanitarian relief should be provided to them by the NGOs.

3. Aid will not be used to further a particular political or religious standpoint. In justifying this principle, the Code states that, although NGOs have a right "to espouse particular political or religious opinions, we affirm that assistance will not be dependent on the adherence to those opinions."

Administering a political or religious litmus test to needy people would certainly interfere with relief efforts. If NGOs adopted such a litmus test, they would automatically become the enemy of any warring party who would not wish to pass the test. Aid workers would then become targets.

But consider this. Providing relief regardless of political or religious standpoints tends to diminish or even devalue the political or religious motives for war. In fact, these motives are then implicitly portrayed as the very cause of human suffering. Therefore, the political or religious motives are made base or labelled inhumane compared to the goodness found in the actions of humanitarian relief organizations.

4. We shall endeavour not to act as instruments of government foreign policy. This principle affirms that independence is what will allow NGOs to focus entirely on relief. "We therefore formulate our own policies and implementation strategies and do not seek to implement the policy of any government, except in so far as it coincides with our own independent policy." No intelligence gathering will be done on behalf of any political entity, whether a warring party, an IGO, or a donor government. Only information that serves humanitarian purposes will be gathered.

This principle is designed to prevent an association with any political entity—an association that would subject an NGO to attack by the enemy of that political entity. Having no allies produces no enemies. No enemies would allow NGOs "to serve purposes that are strictly humanitarian."

But consider this. An independent relief strategy will differ from warring strategies. Military offensive operations may be blocked as aid workers physically get in the way. Information on the humanitarian situation may indicate which warring party's criminal action created the suffering. The international community, once so informed, may intervene—calling it humanitarian intervention—to stop the violation of human rights. Humanitarian intervention, whether in Somalia, Haiti, or Bosnia, upsets the natural flow of war and takes it out of the control of the warring parties. Victory becomes elusive. Not being an instrument of government foreign policy makes humanitarian assistance an instrument of humanitarian principles, which, by their nature, are antithetical to civil wars.

5. We shall respect culture and custom. This principle of the Code proclaims: "we will endeavour to respect the culture, structures and customs of the communities and countries we are working in."

This respect for the local culture facilitates relief. The closer NGO aid workers get to the people, the more the people will accept the aid workers and their humanitarian assistance. Novice NGOs, not understanding cultural issues such as gender roles, tribal loyalties, or dietary habits, can and have been tarred by warring parties as enemies of the people.

But again, consider this. NGOs, by getting close to the people, are mobilizing the people to stop the violence and to reduce the effects of wars. They will be draining civilian support for the warring parties because civilians will no longer have to rely on their political leaders for food, shelter, protection, and health care. Civilians can become, in effect, wards of the NGOs. This is especially true of refugees. Camps that are served by aid workers reduce the mass base available to one or more warring parties. As we will review in our case studies in later chapters, this hurts the warring parties in their armed struggle in places such as Rwanda.

6. *We shall attempt to build disaster response on local capacities.* "Where possible, we will strengthen these capacities by employing local staff, purchasing local materials and trading with local companies . . . and cooperate with local government structures where appropriate." Coordination with other relief organizations, including UN agencies, will also be a "high priority."

Relief agencies working together and not at cross-purposes, while integrating with the local economy, will definitely maximize humanitarian assistance. The problems of the NGOs' importing material when the same material is available locally, duplicating services but ignoring others, and forgetting long-term developmental needs, will be minimized by following this principle.

But is this not another principle, if followed, that would organize a people for a better life and not for war? People, materials, and local officials are removed from the war and directed to non-war projects.

7. *Ways shall be found to involve program beneficiaries in the management of relief aid.* "Full community participation," and not the imposition of disaster assistance, makes for "effective relief and lasting rehabilitation."

Yes, indeed. And while NGOs are at it, they siphon off local leaders to focus on social services to the detriment of the war effort. All wars need leaders, and not only for the fighters in the field. Perhaps as important, the civilian sector needs leaders to recruit new fighters for the military, to mobilize the economy to provide war material, and to keep up morale. Strip these civilian leaders from the war and the warring parties will have trouble sustaining themselves.

8. *Relief aid must strive to reduce future vulnerabilities to disaster as well as meet basic needs.* "All relief actions affect the prospects for long-term development." We will also seek, the Code affirms, "to avoid long-term beneficiary dependence upon external aid."

Creating "sustainable lifestyles" certainly creates hope among aid recipients that the future may be brighter. It increases the attractiveness of NGO assistance.

But it does affect the war. Development aid tends to insulate people from war since they can rely on themselves (and NGOs) for basic needs and less on the warring parties, whether governments or rebels. Here again, NGOs, to varying degrees, drive a wedge between civilians and the warring authorities.

9. We hold ourselves accountable to both those we seek to assist and those from whom we accept resources. In support of this principle, the Code advocates "openness and transparency" in reporting NGO activities from both the financial and effectiveness perspectives. The impact of the aid will be monitored, as will "the factors limiting or enhancing that impact." "High standards of professionalism" will be observed in order to minimize waste.

With less playing to the media and more serving the needs of the victims of war, not only will more aid be delivered, but the image of humanitarian relief will remain high. Reports by the NGOs must not sugar-coat the war or proclaim how wonderfully relief efforts are proceeding. This is essential to nurture and keep donors. Showboating and scandals can devastate humanitarian agencies.

But this principle, too, has a political impact. An objective view of war and a report of the difficulties faced in bringing relief to its victims do not enhance the image of war. The reports inevitably become propaganda barrages against the warring parties. The warring parties create the victims. They, along with any fighting peacekeepers, interfere with relief operations to "enemies." No relief operation can end the suffering; to some degree, they are all bound to fail. War cannot be sanitized, and NGOs will remind donors and recipients of this fact repeatedly. War is bad.

10. In our information, publicity and advertising activities, we shall recognize disaster victims as dignified human beings, not objects of pity. Here again, NGOs "shall portray an objective image of the disaster situation where the capacities and aspirations of disaster victims are highlighted, and not just their vulnerabilities and fears." Cooperation with the media will be managed "in order to enhance public response" for "maximizing overall relief assistance."

This also reads as another reaffirmation of a non-political stance. The impact of following this principle, however, is no less than an assault on the very foundation of war. Wars are fought against enemies, despicable enemies, enemies that one cannot reason with. No war has been fought without somehow demonizing the evil, degenerate, immoral or aggressive people on the other side. In war, they must be worthy of death. If the NGOs cooperate with the media and refer to war victims as "dignified human beings, not objects of pity," then any killing in war becomes close to criminal. The victims get human faces. "Victims"—the very term denotes that something wrong has happened to them, something undeserved, and something requiring a "public response." The media will be used to "enhance" that response, "maximizing overall relief assistance." In effect, the Code calls for NGOs to use the media to increase third-party interference in war in order to make a more congenial working environment for humanitarian assistance.

THE WORKING ENVIRONMENT: ANNEXES

The ICRC and NGO representatives meeting in Birmingham in 1993 did not stop after adopting the Code of Conduct. They also approved:

> [S]ome indicative guidelines which describe the working environment we would like to see created by donor governments, host governments and the intergovernmental organizations—principally the agencies of the United Nations—in order to facilitate the effective participation of NGHAs (non-governmental humanitarian agencies—another name for NGOs engaged in international humanitarian efforts) in disaster response.

The guidelines continue the theme of getting all the participants in wars to behave in ways to ameliorate the disaster of war. Wars should present a "working environment" congenial to NGOs! The guidelines seem to reverse ICRC President Sommaruga's statement that "humanitarian action must be in parallel with political action." Instead, political action must be in parallel with humanitarian action! The action of both sets of third-party interveners in war, the political/military ones and the humanitarian ones, will then have reinforcing effects. It is the international community ganging up on the institution of war.

Annex I presents five guidelines, five "[r]ecommendations to the governments of disaster-affected countries." Disaster-affected countries are where wars are fought. Their governments are warring parties. Again, the guidelines are presented only in terms of facilitating relief. Their political effects are unstated, yet political effects they do have.

1. Governments should recognize and respect the independent, humanitarian, and impartial actions of NGHAs. This guideline makes NGOs free to protect and provide assistance to the wounded, prisoners, detainees, refugees, and civilians, providing aid as the needs of the victims of war dictate. NGOs will operate beyond the control of the warring parties.

The fox allows the guard dog into the chicken coop.

2. Host governments should facilitate rapid access to disaster victims for NGHAs. The ICRC wants host governments to waive "requirements for transit, entry and exit visas, or arranging for these to be rapidly granted" and to "grant over-flight permission and landing rights for aircraft transporting international relief supplies and personnel, for the duration of the emergency relief phase."

In effect, one warring party is to cooperate actively in facilitating outside intervention in the war, including relief operations to its enemy. Killing, wounding, expelling, and destroying are to be quickly prevented or remedied.

3. Governments should facilitate the timely flow of relief goods and information during disasters. No duties or licensing or port charges shall be imposed on NGOs for bringing in relief supplies and equipment. Radio frequencies should be designated for use by relief agencies.

War is not a pleasant thing to see. Host governments, though, are to facilitate communications to let all the world see the ugly residue of war—the grieving, the homeless, the injured, the dead. All wars are filled with atrocities, and the ICRC, Médecins sans Frontières, and other NGOs have made an art out of whistleblowing on the perpetrators of these atrocities.[11] And in contrast to the bloodthirsty fighters, in their midst are the well-equipped humanitarian relief workers acting to undo the dreadful conditions created by the fighters.

4. Governments should seek to provide a coordinated disaster information and planning service. "The overall planning and coordination of relief efforts is ultimately the responsibility of the host government. Planning and coordination can be greatly enhanced if NGHAs are provided with information on relief needs and government systems for planning and implementing relief efforts as well as information on potential security risks they may encounter. Governments are urged to provide such information to NGHAs."

Making governments responsible for relief tends to put those warring parties somewhat at cross-purposes with their wars. They must take measures to clean up the mess they and their enemies have created. Perhaps the most antiwar recommendation is the request for governments to provide information "on potential security risks" NGOs may encounter. No warring party would do that; it would reveal its future combat operations or where the enemy is expected to attack. Strategic and tactical moves are hidden in war so that the enemy cannot prepare to counteract them. Surprise allows superior force to be applied on an objective. No wonder these guidelines are recommendations. Annex I, in so many words, recognizes the difficulty in making them legally binding, "although this may be a goal to work towards in the future."

5. Disaster relief in the event of armed conflict. "In the event of armed conflict, relief actions are governed by the relevant provisions of international humanitarian law."

The Code and this Annex have had war and war relief principally in mind all along. Not making a clear distinction between natural disasters and war disasters makes the largely non-political natural-disaster relief a mask for legit-

An ICRC Delegate Registers Detainees in Bosnia.
In a prison in Kaonik, we see the faces of detainees who are about to be registered, which gives them great comfort. If they disappear, their names will be passed on to UN officials. In light of the International War Crimes Tribunal in The Hague, those who detain people might think twice about "disappearing" people. The ICRC is legally mandated to visit detainees and POWs, and this serves to deter savagery. *(Photo courtesy of ICRC, by Ana Feric)*

imizing the political-war relief. When this guideline refers to armed conflict, the buck is passed to politically sanitized international law.

The ICRC, as noted in the chapter 1, has been in the forefront of formulating, gaining adherence to, and disseminating international humanitarian law. One area of this body of law not yet analyzed adds further proof to the antiwar political effects of the ICRC. That area is the banning of weapons.

The ICRC's first efforts in banning certain weapons arose out of World War I. Poison gas proved to be an effective weapon in trench warfare, killing and incapacitating hundreds of thousands of troops on both sides. Effective countermeasures—masks and protective clothing—reduced the potency of the weapon, but new, more lethal gasses were being researched and developed. In 1918, even as the Great War raged, the ICRC called for the banning of chemical weapons. With the support of the victors, the Geneva Protocol of June 17, 1925, prohibited "the use in war of asphyxiating toxic and similar gasses and of bacteriological means."[12]

The ICRC's most recent efforts sought a comprehensive ban on the production, transfer, stockpiling, and use of anti-personnel land mines, a ban on laser weapons designed to blind, and restrictions on the arms trade. The ICRC produced studies that showed that over 20,000 people are killed or maimed annually from land mines, most of them civilians. More than 100 million land mines are buried in 64 countries. It helped organize over 350 human rights groups to come together as the International Campaign to Ban Land Mines. It successfully lobbied Secretary-General Boutros Boutros-Ghali to publicize the issue and to call an international conference to add new rules to the 1980 Convention on Inhumane Weapons.[13] Asked why the ICRC favored a total ban on anti-personnel mines, ICRC President Cornelio Sommaruga replied:

> We were pushed by our delegates in the field. I have medical delegates who are operating again and again and again on victims of these antipersonnel mines. In our 10 war-surgery hospitals, in many other hospitals where we have surgical teams which are going and helping, what do we see? Children, women, elderly people who are not combatants who are hurt by these anti-personnel mines.[14]

The UN-sponsored conference convened in Vienna in September 1995 and after three weeks of discussion failed to reach a consensus. Delegates from China, Russia, India, Pakistan, and Iran raised objections. Their main objection was that land mines are cheap, effective weapons. One military source indicated that "nothing does the job for less."[15] Anti-personnel mines disable more than kill, which hurts troop morale and ties up medical personnel and resources. They are excellent weapons for defense—delaying, channeling, and frustrating an enemy. Even Red Cross publications call anti-personnel mines

> "perfect soldiers": they are always at their posts, never sleep and never miss, even when hostilities have long since ended. But they are also perverse and totally ignorant of international humanitarian law.[16]

And anti-personnel mines don't have to be fed or paid. To ban them is to take away a weapon that can hasten victory or prevent defeat. War would suffer as an institution to settle conflicts, an effect similar to that of banning poison gas. And that is precisely the unspoken mission of the ICRC, joining the public mission, which is to protect innocent victims who every day lose lives and limbs to this dreaded weapon. The UN, with full NGO backing, decided to reconvene the conference in 1996. What emerged from this conference in Geneva was a tepid protocol that failed to ban anti-personnel mines, although a provision was approved that required governments in a decade to employ remotely only mines that self-deactivated within four months.

The Vienna Conference on the 1980 UN Weapons Convention made more progress with laser weapons designed to blind. A protocol was approved, banning such weapons. One of the U.S. delegates to the Conference, not a great fan of the ICRC, told me "It was easy to agree to ban a weapon that has not yet been developed." Technically, this is correct. Lasers designed for ranging and targeting weapons, since they were not designed to blind the enemy, are excluded from the protocol, but of course could be used to blind. The same U.S. official expressed a mixture of respect and disapproval by saying that the ICRC spent over $10 million on its recent anti-weapons campaign, and that the U.S. government, his employer, is the largest contributor to the ICRC budget. "We're paying them to lobby us."

The 1995 International Conference indicated that the next banning campaign will target the trade and availability of small arms.

Since war is by nature inhumane, to ban all inhumane weapons is to ban war. Nothing would please the ICRC more.

Annex II presents "Recommendations to Donor Governments." Keep in mind that the principal governments that donate to the ICRC are those in North America, Western Europe, and Japan. These governments are the ones that approve, fund, and provide the personnel for most UN peacekeeping operations in internal wars. Again, we see the guidelines performing two functions: making humanitarian assistance effective and making war ineffective.

1. Donor governments should recognize and respect the independent, humanitarian, and impartial actions of the NGHAs. "Donor governments should not use NGHAs to further any political or ideological aim."

In one sense, this guideline says that if the major powers have political and ideological aims, they must pursue them themselves. Do not use NGHAs. Rather, join them in the right political and ideological aim—stopping war. If that is not possible, let them alone in providing relief.

2. Donor governments should provide funding with a guarantee of operational independence. And the more funding the better in advancing "a spirit of humanity!" We might add that NGO "operational independence" will dampen the spirit of war.

3. Donor governments should use their good offices to assist NGHAs in obtaining access to disaster victims. "Donor governments should recognize the importance of accepting a degree of responsibility for the security and freedom of access of NGHA staff to disaster sites. They should be prepared to exercise diplomacy with host governments on such issues if necessary."

This is a barely concealed plea for major states to intervene in internal and regional wars and then adopt an antiwar posture. To advocate the access of humanitarian agencies to the battlefields implies that access is often denied, that atrocities occur there, that the nature of the war needs to be changed.

Chapter 4 will show how important the ICRC is, along with other NGOs, in attracting third-party humanitarian, military, and diplomatic intervention in wars. Perhaps more than any other function, this expansion of the number of participants who do not have war aims or interests in who wins the war is the most antiwar activity of the ICRC. With UN-authorized peacekeepers involved, the war becomes uncontrollable by the combatants. A balance of power between belligerents can never be stabilized by the belligerents themselves. Peacekeepers are now more frequently prepared to become peace enforcers and to use combat. The ICRC uses the UN; it lobbies its members, and it has guidelines for the UN as well.

Annex III presents "Recommendations to Intergovernmental Organizations."

1. IGOs should recognize NGHAs, local and foreign, as valuable partners. "NGHAs are willing to work with UN and other intergovernmental agencies to effect better disaster response. They do so in a spirit of partnership which respects the integrity and independence of all partners. Intergovernmental agencies must respect the independence and impartiality of the NGHAs. NGHAs should be consulted by UN agencies in the preparation of relief plans."

This recommendation practically invites UN humanitarian intervention "to effect better disaster response."

2. IGOs should assist host governments in providing an overall coordinating framework for international and local disaster relief. "NGHAs do not usually have the mandate to provide the overall coordination framework for disasters which require an international response. This responsibility falls to the host government and the relevant United Nations authorities."

The 188 states that have ratified the four 1949 Geneva Conventions are responsible for providing an "overall coordinating framework" for both domestic relief agencies and the ICRC. The United Nations, however, is not a state and cannot sign a treaty. What responsibility the UN has comes from the resolutions passed by its members. This recommendation seems to create an obligation or responsibility for the UN to intervene if the disaster is big enough to "require an international response." The ICRC does not want wars to be isolated where "war might make right." Supposedly, the UN, as its Charter proclaims, operates on the basis of principles designed "to save succeeding generations from the scourge of war" and "to reaffirm faith in fundamental human rights." Injecting the UN into wars injects these anti-war principles into the struggle.

3. IGOs should extend security protection provided for UN agencies to NGHAs. "Where security services are provided for intergovernmental organizations, this service should be extended to their operational NGHA partners on request."

This recommended guideline calls for the protection of NGO workers in highly dangerous war zones. The internal wars in which the ICRC now operates tend to be exceptionally brutal. Governments train their regular troops to follow orders, maintain discipline, and obey the laws of war. International wars, therefore, follow international law to the degree the belligerent governments command. Internal wars, however, find at least one side whose fighters are not trained in this fashion. Guerrillas, militias, and terrorists regularly violate the laws of war, from not wearing uniforms to raping and killing civilians. When, as in Somalia, Liberia, and Afghanistan, countries that have experienced conditions in which no institutionalized government exists, wars can get brutal beyond belief. So what the ICRC is saying is that in really brutal wars the UN should be involved, be sufficiently armed to provide security for its own relief agencies, and also be powerful enough to extend security to the NGOs in the field. Armed peacekeepers, as our case studies will show, eventually become participants in many wars. They always shape, even author, the agreement that ends the war. The only exceptions, of course, arise when the peacekeepers stay away—as they did in Sudan, Afghanistan, Sri Lanka, and Chechnya—or cut and run—as they did in Lebanon in 1984 and Somalia in 1994. In Lebanon and Somalia, the warring parties did not want to settle and were strong enough to bloody the peacekeepers. Inviting in the IGOs and keeping them in the conflict area inevitably messes up the war; the warring parties eventually are induced to settle.

4. IGOs should provide NGHAs with the same access to relevant information as is granted to UN agencies. No cover ups. No trying to keep the horror of war from being accurately described, via the NGOs' extensive media contacts, to the world.

The ICRC calls the Code of Conduct and its Annexes "principles," "guidelines" and "rules." It does not seek to make them law by convention, which would be difficult because of problems of interpretation and the likely opposition of states with a military tradition. Instead, the Code and Annexes are "voluntary and self-policing."[17] The hope is to have the Code and Annexes eventually become established as international customary law.

THE CODE OF CONDUCT AND ANNEXES UNDERMINE WAR'S EFFECTIVENESS

We can strengthen our argument that the Code and Annexes undermine wars by examining what has made war so popular and so frequent in human history.

Anthropologist Margaret Mead's familiar declaration that "war is only an invention" is predicated on an understanding of humans' propensity to create structures and processes to handle persistent problems efficiently.[18] Without great elaboration, the structure of war embodies the creation of armed forces and providing them with the stuff of war—munitions, logistics, transportation, communication, operational plans, worthy motives, and popular support. The process is combat and threats of further combat, both of which are targeted to destroy the enemy's armed forces, infrastructure, and popular support.

The institution of war is employed to resolve conflicts. The equally familiar declaration by Carl von Clausewitz that "war is the continuation of policy by other means" puts war into the realm of the rational. Whether offensive or defensive, it is a means, a strategy for achieving political goals. What makes war rational as an institution or strategy?

War as rational activity depends upon the stakes involved, the options available, and the expectations of costs and benefits in the minds of political leaders.

States and other political groups (revolutionaries or separatists) go to war if their *vital interests are at stake,* their leaders perceive that there is *no other least-costly option* to protect or advance those interests, and there is *an expectation of winning at bearable costs.*

Vital interests involve the security of the state or group—its people and their beliefs, its territory, and its government. Big wars have big stakes among big powers, as was true in World War II. Lesser wars still have big stakes, relatively, to lesser powers—for example, in the Falklands/Malvinas War. Historically, wars are rarely fought over trivia. War becomes the only option when the enemy will not negotiate and accept the security of one's state or group. There must also be the expectation of prevailing or winning, which depends upon power relations. A state or group will war if it expects to have greater power than the enemy. And the main ingredients of power have been military capability, good leadership, popular support, and dependable allies. Historically, no party has gone to war expecting to lose.

Thus, the process of war for both sides involves enhancing the ingredients of one's power and lessening that of the enemy to the point that the enemy's will to resist is broken. Except for a Carthaginian peace (that is, the complete destruction of the enemy), wars are designed to end with a treaty that confirms the goals of the victors. War has traditionally been a way to get the whole loaf, decisively.

Notice how following the Code of Conduct and Annexes corrupts the structure and process of war and undermines its effectiveness as an institution to resolve conflicts.

Overall, it injects other parties not tied to the war into the war. These other parties include UN-sponsored diplomats, UN humanitarian relief agencies, UN-approved peacekeepers, the armed forces or hegemonic powers, and NGO relief agencies. They:

1. Create a viable diplomatic forum for a settlement, thereby opening up a new, negotiation option for dealing with the conflict. This has happened in all post–Cold War conflicts with the possible exceptions of those in Sudan, Sri Lanka, Indonesia/East Timor, and Turkey/Kurdistan (some third-party diplomatic efforts were even made in these wars). The ICRC would like to remove these exceptions.

2. Modify the stakes by creating half loaves that are better than a stalemate; that is, create specific terms for a settlement that satisfy some of the basic interests of the warring parties. The United States, for example, even wrote a sample constitution for Bosnia for the UN-approved negotiations in Dayton among Bosnia, Croatia, and Yugoslavia.

3. Promote stalemates by upsetting power relations through intervention with peacekeeping or peace enforcement troops and through sanctions that limit the import of munitions. To illustrate, no side in the wars in the former Yugoslavia could calculate the eventual attainment of superiority. Power was made fluid by the twists and turns of NATO peace enforcers—bombing here, blocking weapon resupply there, winking at weapon resupply at other places.

4. Constrain the use of military force by protecting civilians, banning weapons (for example, through "no-fly zones"), and deterring attacks on civilians and POWs and thereby prevent the breaking of the enemy's will to resist. Surrender has become virtually obsolete in current wars. With the UN and

NGOs providing sustenance and protection, why give up? The ICRC, by making surrender obsolete, would like to make war obsolete.

5. Isolate the conflict by stopping the warring parties from gaining allies through UN Security Council resolutions, blockades, and troop deployments (as was done, for example, in Macedonia).

6. Weaken popular support by breaking down stereotypes, tarnishing the motives for war, getting civilians from both sides together, and raising humanitarian principles. This is a recipe for war weariness, and war weariness diminishes the popular support for war and for the leaders of the warring parties.

As Margaret Mead pointed out, institutions invented by humans can be abandoned by humans, as was achieved with slavery and dueling. Like slavery and dueling, institutions can be discarded if something less costly and more effective takes their place. That is: 1) if those that try to use the institution of war can be punished by outsiders; 2) if wars fail to resolve problems; and, 3) if the warring parties see other political groups successfully use non-violent means to manage and resolve conflicts.

The guidelines in the Code of Conduct and the three Annexes undermine the institution of war in the above three ways. They are not friendly toward war. Undermining war, above all, depends upon activating the UN and influencing the mandate of its peacekeeping forces, and the ICRC has become proficient at both tasks. How the ICRC operates to expand the number of participants in wars, participants more concerned about stopping wars than about who wins, is the subject of the next chapter.

4

Lobbying the Peacekeepers

The ICRC may be the world's premier lobby. Alone or in tandem with the International Federation, it frequently sponsors symposia, conferences, and meetings. It does so to gain access to leaders of other NGOs, governments, and IGOs. To be a player in "humanitarian politics," and a powerful player at that, knowing political and NGO leaders and developing access to their decision-making processes is absolutely essential.[1] Basically, these meetings bring together experts in the broad areas of international law, peacekeeping, and humanitarian relief. ICRC officials use these meetings to discover the interests of these other players. They probe what new international law they can expect to get approved. They produce thoughtful studies on the problems new law is to address, which they then disseminate freely and widely. Symposia, conferences, and meetings are the favorite venues for formal ICRC lobbying.

Much of the focus of ICRC conferences in the post–Cold War period has been on the ICRC's relations with the more active UN and its members. Officials in Geneva do their political homework. They are all very aware that their twin missions of providing relief to the victims of war and of undermining war depend upon the actions of political leaders. They know that a new international system creates new threats and opportunities for political leaders, and therefore creates new foreign policies. Here is how ICRC President Cornelio Sommaruga prefaces the report of a "Symposium on Humanitarian Action and Peace-Keeping Operations," held in Geneva, on June 22-24, 1994.

> The end of the Cold War raised the legitimate hope that the number of conflicts would drastically diminish. However, although there did follow an easing of tensions between the nuclear powers and a significant slowing down of the arms race, violent conflicts have since flared up in different parts of the world, claiming millions of victims, most of whom are, tragically, civilians and children.
>
> At the same time, the changed structure of contemporary international politics has definitely enhanced the possibilities for remedial action by the United Nations, and in particular by the Security Council, which has been freed from the despairing paralysis it suffered during the Cold War era. Consequently, recent years have witnessed a dramatic increase in the number, and at times scale, of operations

> established by the United Nations, involving the use of military means and entrusted with various tasks such as keeping the peace, protecting the population from the effects of hostilities, or establishing a secure environment to allow distribution of humanitarian assistance.[2]

This growing UN role affects the ICRC's twin missions directly. First, UN intervention in wars adds its relief organizations—the World Food Program (WFP), UNICEF, and the UN High Commissioner for Refugees (UNHCR), principally—to those of the ICRC and other NGOs in providing humanitarian assistance. This raises enormous coordination problems. Second, the UN's frequent use of "military means," or peace enforcement, makes the UN an actual participant in these wars. This, too, affects relief operations. Peace enforcement is not neutral. Combat never can be neutral because an enemy is required.* UN combat ties both its and, by association, the NGO's relief agencies to one side, thus dashing the neutrality of the ICRC. This raises questions about access to all victims and about the protection of relief workers in the field. Above all, it raises the question of controlling UN combat via international humanitarian law so that wars are lessened in intensity and quickly resolved. As I argued in chapter 3, the UN intervention in wars is key for making wars ineffective and therefore dysfunctional.

Any time party A depends upon the behavior of party B for the conduct of A's missions, you will see party A devoting a massive effort to affect party B's behavior. The ICRC is party A and the UN is party B.

Confirmation that the ICRC leadership now engages in a massive and coordinated lobbying effort can be found in the agenda of the 1995 International Conference. The Conference divided into two study groups: Commissions I and II. The Commissions were charged with deliberating the two issues created by more robust UN operations in war.[3] Commission I dealt with "War Victims and Respect for International Humanitarian Law." This is a euphemism for strengthening and disseminating international humanitarian law in order to reduce the violence and suffering caused by war. Commission II dealt with "Humanitarian Values and Response to Crises," which is shorthand for how the Red Cross Movement and other NGOs can relate to UN forces and agencies once they are involved in wars so that the ICRC's twin missions can be fulfilled.

---

*Theoretically, UN-mandated forces could target all the sides in an internal or regional war. They came close in Bosnia when, after air strikes and artillery bombardment of the Bosnian Serbs in the summer of 1995, Croat and Bosnian Muslim forces—after their subsequent successful offensive—kept fighting beyond the cease-fire deadline. United Nations-sanctioned NATO leaders threatened possible military action, and that was sufficient to stop the Muslim-Croat offensive.

Six ICRC lobbying objectives targeted at the UN and its members can be clearly identified:

1. Formalize a commitment for UN members to get involved in internal wars and to provide sufficient resources to make the intervention powerful.
2. Expand UN humanitarian efforts to reduce suffering while preserving the independence of NGO relief operations.
3. Reduce the use of force by UN authorized peacekeepers as these peacekeepers likewise induce the warring parties to reduce their use of force and settle the conflict.
4. Develop operational procedures that the UN accepts for dealing with NGOs.
5. Energize the UN and the entire international community to engage in preventive diplomacy and deterrence to prevent potential warring parties from reaching that potential.
6. Engage the UN process in codifying an expanded body of international humanitarian law that will support objectives 1-5 while approving plans for the ICRC to train diplomats and troops to understand and carry out that law.

Each will be considered in turn.

Objective 1. UN Commitment to Intervention. The ICRC chairman of the 1994 Symposium on Humanitarian Action and Peace-keeping Operations, Yves Sandoz, offered this conclusion to the three-day meeting. The United Nations, the NGOs, and the ICRC were all necessary to ameliorate wars, Sandoz prefaced, and "in the long run, they all needed each other."[4]

> One of the recurring themes in the discussions was indeed the need for clarity in order to increase efficiency and, whenever necessary, to ensure better collaboration. However, clarity was required at a host of levels. The international community had to state clearly what it considered unacceptable, be it massive violations of international humanitarian law, human rights or situations of total anarchy where nothing but chaos and violence prevailed. Moreover, it had not only to make clear that it refused to accept such situations, but also to adopt the means necessary to put an end to them. And these situations demanded far more than what humanitarian assistance could provide, for what was needed was not just sacks of flour or medical kits but an appropriate response, in terms of personnel and equipment, to halt the catastrophe.[5]

What Sandoz pleaded for was nothing less than a standard by which the UN would intervene in wars and then employ sufficient power to end violations of international humanitarian law and human rights, and thus end situations of chaos and violence. This standard would serve to shape the national interests of member states and thus compel them to adopt interventionist foreign policies towards civil wars under the aegis of the UN. Of the 40 non-ICRC participants at the Symposium, 12 were associated with the UN, including 4 officials from the UNHCR and 4 engaged in military peacekeeping operations. The majority of the remaining 38 participants from governments or NGOs and the 11 ICRC representatives could be counted upon to generate a consensus for formalizing UN intervention to undertake war relief and war termination.

To halt "the catastrophe" of war, the ICRC prefers a diplomatic solution over UN peace enforcement (combat). Jean de Courten, ICRC Director of Operations, made this clear in his presentation at the Symposium. UN "military measures," even "where necessary" as "an integral part of a political strategy for the settlement of conflicts," produce reservations by the ICRC because these military measures lead to "the politicization of humanitarian action."[6] Mr. de Courten illustrated the general ICRC position by referring to the civil war in Somalia.

> In December 1991, the President of the ICRC had written to the United Nations Secretary-General Pérez de Cuéllar, not however for the purpose of requesting armed protection or intervention, but to describe the situation prevailing in Somalia, a kind of early-warning of the impending catastrophe and a call for the United Nations to join the ICRC in the field to cope with the dramatic situation. This was rapidly answered by the World Food Programme. Armed intervention was decided upon much later, and certainly not at the demand of the ICRC.[7]

Mr. de Courten's remarks reveal one of many ICRC lobbying efforts to engage politically the UN in wars. One function of all lobbying groups is to be a watchdog or whistle blower (the ICRC prefers to use the less-political term "early warning" to describe this activity). Knowing what will stimulate the powerful players who can remedy the situation provides lobbies with the ability to tailor that stimulus. "Describing the situation" in a certain way will elicit the proper response, and for the UN, that response means that UN forces will engage in peace operations, provide humanitarian relief, and facilitate conflict resolution in internal wars. Since these wars, compared to international wars, are fought with weaker sides using less powerful weapons, a UN presence will loom large. It will usually be backed by the foreign policies of the major powers, which can use economic sanctions, propaganda, war-crimes-tribunal threats, diplomatic mediation, and even (if regretfully necessary) peace enforcement operations to inject enormous outside power that is hostile to these little wars.

In a follow-up effort to formalize what will bring about UN peace operations, the ICRC did what it always does. It began the process of refining international humanitarian law. First, the ICRC convened a "Meeting of Experts on the Applicability of International Humanitarian Law to United Nations Forces" in Geneva, March 6-8, 1995. The participants, including Lamin Sise of the UN's Department of Peace-keeping Operations, were the very ones who would be called upon to create the norms for triggering UN peace operations/humanitarian intervention and who would help to define what provisions of international humanitarian law would apply to UN forces. The meeting was predicated on the assumption that the UN would be involved in local and regional wars.[8]

We will look at some of the recommendations in the next section, but for now, the meeting confirmed that the ICRC sees international humanitarian law, supported by governments, as the key to effective UN peacekeeping and peace enforcement.

The 1995 International Conference of the Red Cross and Red Crescent, especially Commission II, devoted considerable effort to create the expectation for UN humanitarian intervention and conflict resolution. Item 2 in Commission II's agenda, titled "Principles and Response in International Humanitarian Assistance and Protection," included the pledge that "[t]he International Federation and the ICRC intend to pursue the on-going co-operation with the UN Inter-Agency Standing Committee" (made up of humanitarian, development, and peacekeeping/peace-making agencies).[9] The agenda then reviewed the "linkages between humanitarian actions and political, economic, and military action." The candidness of the statement makes it worth quoting at length.

> Dilemmas abound in today's humanitarian agenda. To feed refugees or help seek out perpetrators of human rights violations? To provide humanitarian assistance or lobby for political action? To seek a high profile in the media, or get on with the work anonymously? Dealing with these dilemmas is part and parcel of humanitarian work today.
>
> While humanitarian action helps meet basic needs and alleviates suffering, it cannot cure the root causes of suffering. No crisis can be solved without political action. Emergency humanitarian aid alone can do no more than temporarily alleviate the acute symptoms of an endemic "disease." The problems of Somalia, Afghanistan, Azerbaijan, Former Yugoslavia or Rwanda cannot be solved solely with humanitarian aid.
>
> This is not to detract from the need for effective humanitarian action. The International Federation and the ICRC remain firmly convinced that a special place must be reserved for humanitarian action, allowing the suffering victims to be reached without delay, independent of any political considerations.

> Sadly, the term "humanitarian" is often used today in a sense quite remote from its original meaning, which is closely tied to the prevention and alleviation of suffering. Creating space for true humanitarian work does not imply isolation or political naivety, quite the opposite. Political action aimed at mobilizing States and the United Nations to ensure greater respect for humanitarian norms and international humanitarian law is essential for the conduct of humanitarian activities. . .
>
> The pressure exerted on governments, as displayed by present day media coverage, has created a political demand for high-profile action. Such an action can lead governments to lose sight of broader needs and to avoid or postpone necessary political or even military decisions. We must reiterate that humanitarian action is no substitute for these decisions.[10]

The working paper for this agenda item, after this revealing passage linking humanitarian and political/military action, is quick to point out that a "clear operational distinction has to be drawn between military and humanitarian action."[11] The dilemma, expressed at the beginning of the above passage, remains. Politics can be poison for ICRC's main mission but must be the antidote for its unspoken one. The UN must be brought into internal wars but in such a way as to not politically taint humanitarian agencies.

This makes the UNHCR one of the ICRC's favorite agencies. It reinforces the perceived ICRC-UN connection in terms of humanitarian relief.

In a plenary session, the International Conference passed Commission II's resolution by consensus (as indicated by applause; no formal votes were taken) that "*calls upon* States: *to redouble* their efforts in the resolution and prevention of conflicts, [in] peace-keeping, disaster preparedness and disaster mitigation, to which the humanitarian work of the [International Red Cross and Red Crescent] Movement acts as a necessary complement."[12] States would perform these tasks primarily through the UN. And the ICRC is the only NGO accredited to the UN, allowing it to voice its opinion directly and giving it access to meetings of the General Assembly and committees.

Daniel Augstburger, a delegate and information officer at the ICRC's UN observer mission, is not reluctant to refer to his New York office as a "humanitarian lobby." As an example, he related that the ICRC "was screaming at the Security Council about the UN's failure to move" in Rwanda. Indirect lobbying comes when "we let the media know what's going on."[13] Pierre Gassman, also a member of the Public Information Directorate, speaking at a journalist conference meeting simultaneously with the International Conference in Geneva, presented a similar view of ICRC lobbying. Gassman, however, was less impressed with the power of the media.

> We don't have the power to prevent war; we give aid. We would like to prevent war, and preventive diplomacy is one part of our job—a stepping stone to get others involved. It is best done without the media; we talk to international organizations and governments.[14]

Getting the international community involved in the politics of internal wars is key for achieving the unspoken mission.

Objective 2. More UN Humanitarian Action to Reduce Suffering and Less Interference with NGO Operations. The ICRC believes that the victims of war have *a right* to humanitarian assistance under the Geneva Conventions and Protocols. Furthermore, the "high contracting parties" to the Conventions have *a duty* "individually, collectively and through the United Nations, [to] adopt a host of measures to pressure a Party to a conflict into accepting delivery of humanitarian assistance."[15] It follows, then, that the more the UN is subject to international humanitarian law, the more it is obliged to act and act vigorously to reduce the suffering in all wars.

But all wars are not equal. The report of one working group at the 1994 Symposium made this clear. "The Geneva Conventions and the additional Protocols, as they stand, were seen as intended to strike a balance between military necessity and humanity from an interstate conflict point of view, wherein military victory was a predominant factor."[16] Recent wars are largely internal, not "interstate." "The nature of wars has changed," said ICRC Advisor to the Armed Forces Major-General Louis Geiger at the Symposium, "to the point where 90% of those directly affected by fighting are not combatants, as in the past, but civilians."[17] Civilians, of course, are to be protected and their suffering relieved under international humanitarian law. Thus the law demands *greater action* by the signatories to the Conventions and additional Protocols. "Everyone agrees that something must be done," General Geiger logically concluded.

The something he had in mind was a greater UN role, supported by its members, in reducing the suffering caused by the new "civilian" wars. Military victory should not be the "predominant factor," as it is in interstate wars.

General Geiger presented "five principles or humanitarian rules governing the conduct of hostilities in non-international armed conflicts" in his comments at the Symposium. These rules would apply both to the warring parties and to UN peacekeepers.

1. The obligation to distinguish between combatants and civilians, and the ensuing prohibition of indiscriminate attacks;
2. The immunity of the civilian population, and the prohibition, in particular, of attacks on the civilian population as such or against civilians;

3. The prohibition of the causing of superfluous injury or unnecessary suffering and, in particular, of the use of means of warfare that unnecessarily aggravate the suffering of the wounded or render their death inevitable;
4. The prohibition of recourse to perfidy, to kill, wound or capture an adversary by abusing his good faith; and
5. The obligation to respect and protect medical and religious personnel, medical units, and means of transport.[18]

One can only wonder, if 90 percent of those affected by modern wars are civilians, and if civilians, according to international law, are out of bounds to attacks, then 90 percent of wars must be virtually illegal! More realistically, protecting civilians in internal wars removes the object of the war from the war itself. The war becomes meaningless. Divorced from political objectives (which depend on the protection and loyalty of one's own civilians and the destruction or removal of the enemy's civilians), the warring parties would only be in the business of senselessly killing each other.

The antiwar posture of the ICRC, while *never* openly stated so as to abide by its principles of neutrality, impartiality, and independence, is its central tenet with every regard to today's internal wars.

Typical of the international lobby, the ICRC's efforts to expand UN humanitarian relief, while "maintaining total independence of decision and action" for itself, include very specific recommendations to the UN on the rules to follow.[19] At the International Conference, Commission I's resolution, presented to the plenary session for approval, urged states to increase efforts "to bring before courts and punish war criminals" and "to establish permanently an international criminal court." The resolution advocated a list of 62 state obligations "to repress violations of international humanitarian law."[20] Civilians are to be protected; sexual attacks on women are to be condemned; children are not to be recruited for combat; families are to be traced and reunited; civilian starvation and denial of water are to be prevented; and anti-personnel mines and blinding weapons are to be banned.

Another resolution from Commission II:

> *Calls* upon states to recognize the need for the Movement to maintain a clear separation between its humanitarian actions, on one hand, and actions of political, military or economic nature carried out by governments, intergovernmental bodies and other agencies during humanitarian crises, on the other hand, bearing in mind the need for the Movement to maintain, in its humanitarian work, its independence, impartiality and neutrality.[21]

Successful lobbies make it easy for political authorities to say "yes" to what the lobbies want, and having the text of what they want be clear, tested, and refined by extensive deliberations makes saying yes easier.

Objective 3. Reduce the use of force by UN-authorized peacekeepers as these peacekeepers likewise induce the warring parties to reduce their use of force and settle the conflict. The ICRC's main lobbying strategy here is to get UN officials to accede to the principle that UN-sponsored military forces are subject to international humanitarian law even though the UN is not a state and therefore cannot sign the four Geneva Conventions and Additional Protocols. This was the full thrust of the ICRC-sponsored conference in 1995: "Meeting of Experts on the Applicability of International Humanitarian Law to United Nations Forces." At the meeting, the ICRC Legal Division listed the specific categories of international humanitarian law that it thought applied to the UN. In the "law governing the conduct of hostilities and other situations involving the use of force" were provisions that limited warfare to combatants and not civilians, limited "weapons or methods of warfare which needlessly aggravate the suffering of persons place *hors de combat,*" and prohibited attacks on "objects indispensable to the survival of the civilian population, such as foodstuffs, crops, livestock and drinking water installations and supplies."[22] It then asked the question: "Are there any rules on the conduct of hostilities that cannot be applied by and to United Nations Forces?" After a lengthy discussion that focused on the differences between the UN and states under international law, Chairman Marco Sassoli, deputy head of ICRC's Legal Division, was reported to have concluded: "With some doubts on reprisals and concerns on the issue of spies, I believe that the meeting agreed that there were no rules on the conduct of hostilities that could not be applied by and to the United Nations."[23]

The ICRC, persistent as always, did not accept the position of the UN's top Legal Affairs officers at the 1994 Symposium. At that time, the two UN legal officials, Ralph Zacklin and Daphna Shraga, gave their interpretation of the ICRC's position.

> [I]nternational humanitarian law principles, recognized as part of customary international law, are binding upon all States and upon all armed forces present in situations of armed conflicts. What is universally binding upon all States, must also be considered binding upon the universal organization established by States and recognized by them as an independent subject of international law. Consistent with its policy of promoting the universal application of international humanitarian law, irrespective of the legality of the use of force, the legal status of the parties, or legal title to the territory, the

> ICRC advocated the applicability of international humanitarian law principles to United Nations peace-keeping operations, when in the course of self defense they resort to the use of force. Having recognized, however, that the United Nations is not formally a party to the Conventions, and given the nature of the Organization, the applicability of international humanitarian law principles to the Organization would have to be mutatis-mutandis. Thus, rules pertaining to prisoners of war or penal sanctions could not apply, whereas rules pertaining to methods and means of combat, categories of protected persons and respect for recognized signs, would be fully applicable.[24]

Zacklin and Shraga thereby summarized a statement by the ICRC to the General Assembly on November 13, 1992. Their reply, giving the UN position, rejected a portion of the ICRC interpretation. The UN is not party to a conflict when its forces are carrying out their peacekeeping mandate. Nor are they a "power" under the Geneva Conventions. The UN is not a state.

However, in response to ICRC demands that UN forces abide by the Conventions, the UN has included in its "Regulations for the Force" the following: "The Force shall observe the principles and spirit of general international Conventions applicable to the conduct of military personnel." A variation of the same clause was included in the Model Agreement between the UN and Member States contributing personnel and equipment to peacekeeping operations.[25] Step by step, little by little, the ICRC advances its missions via advances in international humanitarian law. Finding the results of the 1994 Symposium less than satisfactory, it convened the 1995 Meeting and got better results.

At its 1995 International Convention, the Red Cross and Red Crescent Movement reiterated two concerns. One pointed out that military action by peacekeepers and peace enforcers, as well as by warring parties themselves, at times illegally increased civilian suffering. With Desert Storm and the wars in the Former Yugoslavia clearly in mind, Commission I's resolution on the "protection of the civilian population in periods of armed conflict" stressed "the prohibition on attacking, destroying, removing or rendering useless any objects indispensable to the survival of the civilian population."[26] In particular, it called upon "parties to conflict to avoid, in their military operations, all acts liable to destroy or damage water sources and systems of water supply, purification and distribution solely or primarily used by civilians."[27] If Red Cross officials had not been so politically discreet, there would have been direct condemnation of the U.S. bombing of Iraq's water, sanitation, and health-care infrastructure in Desert Storm, which produced a high incidence of sickness and death, especially among children. The International Conference also expressed grave concern about the "human consequences of economic sanctions" that increased civilian suffering in wars.[28]

There was some ambivalence here. Commission II recognized that economic sanctions—"unarmed warfare"—may be necessary "when diplomacy fails and war is too drastic or unpalatable a domestic option." However, sanctions "may be indiscriminate, and have a disproportionate effect on the lives of ordinary people." Commission II's working paper defined the problem.

> For the UN Security Council, imposing sanctions exposes a potentially fundamental contradiction in implementing two of its core principles: promoting peace and promoting human rights. Sanctions are intended to deal with the former, but risk undermining the latter during the course of their implementations.
>
> Equally, the principle of proportionality suggests that harm inflicted by sanctions to achieve change should not be out of proportion to the expected good.
>
> Although many States, including the Federal Republic of Yugoslavia, Iraq and Haiti, had already suffered military or economic disruption before sanctions were imposed, a recent study undertaken by the International Federation shows that sanctions appear to hit the most vulnerable and destroy their livelihoods and even their lives thus adding further suffering on top of that inflicted by war.
>
> Since sanctions as instruments of international will are unlikely to be abandoned, the issue of whether such sanctions should be allowed free rein or, like warfare, operate within prescribed limits becomes critically important.[29]

The working paper then stresses that, as with international humanitarian law, the end does not justify the means. The survival of the civilian population must be assured. The UN Security Council, the paper suggests, can take three steps as "a possible way forward." It can:

> [First,] assess the impact of sanctions upon the most vulnerable before they are applied and while in force. Second, [design] sanctions procedures to allow the delivery of necessary humanitarian supplies, such as food and medicines, and ensure access to aid by those in need. Third, [streamline] sanctions procedures to allow humanitarian assistance by designated organizations, such as UN humanitarian agencies and the International Red Cross and Red Crescent organizations.[30]

Sanctions, the paper concludes, should have "political or economic impact . . . upon those in power . . . The Movement urges governments to consider ways of implementing these suggestions." Again, this is full-blown lobbying of the 180 or so government representatives at the International Conference. At the Conference center, huge pictures of women, children, and aged people suffering from the consequence of war adorned the vestibule, as did pictures of Red Cross relief operations. The protection of civilians dominated the discourse.

Objective 4. Develop operational procedures that the UN accepts in its relations with NGOs. Much of this lobbying activity has been reviewed in our analysis of Annex III to the Code of Conduct (see pages 48-50). Key phrases include: "spirit of partnership which respects the integrity and independence of all partners," "consulted by UN agencies," "effective coordination," "extend security protection," and "share information."

The more the UN and NGOs are in synch, the more the essentially antiwar efforts of the NGOs will be reinforced by the UN. Protecting civilians, preventing massacres, relieving suffering, and publicizing the horrors of war do little to advance the cause of war. Combine this with the UN peacekeeping functions of separating combatants, monitoring cease-fires, protecting relief operations, and brokering agreements, and there is a double dose of antiwar activities injected into armed conflicts. A phrase frequently heard at the International Conference portrayed humanitarian work as a necessary complement to conflict resolution and peacekeeping.

Objective 5. Energize the UN and the entire international community to engage in preventive diplomacy and deterrence in order to prevent potential warring parties from reaching that potential. This, too, is a familiar refrain of the ICRC. The 1994 Symposium was larded with laments about violations of law and the suffering imposed upon the victims of war, especially civilians and children. It brings to mind a phrase uttered by an observer of the long civil war in Lebanon: "They will keep fighting each other as long as they hate their enemies more than they love their children."

Hatred is abroad in the world, and it is seemingly more about ethnic, linguistic, and religious differences than it is about clashing national interests. These differences between people are the raw materials of internal wars. The UN has no less a responsibility to prevent these wars than it does wars between members. In his concluding remarks, Symposium Chairman Yves Sandoz is reported to have said that "the international community had an even greater and pressing responsibility to prevent such situations [civil wars] from occurring and to refuse to accept them as fatalities. This would cost less, including in humanitarian terms."[31] One of the working groups at the Symposium agreed that the "[e]mphasis should equally be put on prevention to avoid arriving at desperate situations, and preventive action could include early warning, diplomacy, and coercive measures short of the use of force."[32]

The International Conference also made the plea for the international community "to fight the root causes of conflicts and . . . to find solutions to conflicts."[33] The Movement acknowledged that it knew that humanitarian action alleviated the effects of wars but not the causes. It called on the governments represented at the Conference to address and alter "the economic, political and social processes governing a country" in crisis.[34] We will have more to say about the ICRC's efforts on behalf of preventive diplomacy and socioeconomic development in chapter 7.

Objective 6. Engage the UN in codifying an expanded body of international humanitarian law that will support objectives one to five, while approving plans for the ICRC to train diplomats and troops to understand and carry out that law. ICRC officials have traditionally believed that creating law and then imbuing the diplomats and armed forces in that law is the proper, "non-political" way to minimize the suffering caused by war. Law can help civilize those who war. This was the Red Cross's original modus operandi. It is still a "major issue," according to Yves Sandoz in his concluding remarks to the Symposium.

> The ICRC, within the limits of its capacity, was intensifying its efforts to help contributing States train their contingents in international humanitarian law. However, the primary responsibility fell upon the States themselves. Proper training in this law applicable in situations where they were present was particularly essential for the blue helmets as they were the show window of the United Nations in this domain and were therefore expected to behave in an exemplary fashion.[35]

Two departments of the UN were major targets of this ongoing lobbying effort, Humanitarian Affairs and Peace-keeping Operations. A common curriculum was sought for all troops and another for UN, government, and NGO personnel.

There is a distinct trend towards legalism in the armed forces of most states. Evidence of this can be found in the creation of Institutes of National Security Law,[36] in the increase in the number of international lawyers attached to cabinets, national security councils, and foreign and defense ministries,[37] and in the expanded use of courts martial and special judicial tribunals for the prosecution of indicted war criminals. All this adds up to greater sanctions, largely from states themselves, imposed on those who violate international law. Although—in the case of UN-established tribunals—those indicted are difficult to apprehend, their movements will be forever restricted. Such is the case with General Ratko Mladić, the Bosnian Serb commander, and Radovan Karadžić, the Bosnian Serb political leader, and 50 or so other indicted Serbs, Muslims, and Croats (as of early 1997).

At the International Conference, the ICRC reiterated its duty to promote the Geneva Conventions and the Additional Protocols, especially to the authorities of newly created states where many internal wars now occur. Its newly formed

Advisory Services on International Humanitarian Law offers to train state legal advisors, to help draft domestic legislation incorporating IHL, and to develop statutes on war crimes and their prosecution. The ICRC announced it is preparing "a model manual on the law of armed conflicts for use by armed forces."[38] As usual, it will convene conferences of the Movement and state representatives "to consider general problems regarding the application of international humanitarian law," including a 1996 conference on "unstructured conflicts."[39]

One indication of the potency of ICRC lobbying is the common usage today of the term *international humanitarian law.* When I studied international law in graduate school in the 1960s, this area of law was called *the laws of war.* Most Pentagon officials, resisting the IHL appellation on purpose, still refer to the laws of war. It may be a losing battle. The special relationship of the International Committee of the Red Cross with the Geneva Conventions and the Additional Protocols and its constant reference to them as the basis of IHL have influenced the language of international law textbooks and government officials. What's in a name? One, the laws of war, tends to legitimize war because it calls it lawful. Another, international humanitarian law, portrays law more as a remedy for the deadly disease that war is.

We turn now to case studies in order to illustrate how the twin missions of the ICRC—to provide war relief and to make war dysfunctional—have been pursued and how well they have fared in practice.

5

Operations in the Former Yugoslavia, Rwanda, Sudan, and Afghanistan

The ICRC's unspoken mission of undermining brutal civil wars has the most success when political/military agents of the international community join humanitarian workers in intervening in these wars. Even without the intervention of a political force, ICRC and NGO efforts still soften the horrors of the internal wars and make them more susceptible to resolution.

Ending post–Cold War internal wars, it must be remembered, requires the political/military operations of third parties. The warring parties are too weak, too distrustful of the other side. They need the security guarantees of powerful third parties to end their wars. The more outside forces combine to intervene in wars, the more likely conflict resolution will occur. This is most evident in the former Yugoslavia, but it is also true for the wars in El Salvador, Haiti, Guatemala, the Israeli occupied territories, and Mozambique. It is somewhat less evident in Chechnya, Cambodia, Angola, and Rwanda, although there is some progress being made in these wars toward a diplomatic settlement.

If powerful political/military forces do not intervene or if they intervene weakly or pull out, then the operations of the ICRC and NGOs tend to disrupt the war, weaken the warring parties, and interfere with their power relations. This prolongs the war, often propelling it into slow motion or fits and starts. We see this situation in Liberia, Somalia, Afghanistan, Sudan, Kurdistan, Tajikistan, Kashmir, Sri Lanka, Ngorny Karabakh/Azerbaijan, and East Timor/Indonesia. Making wars dysfunctional eventually makes them ripe for resolution and produces a request by the warring parties for third-party mediation.*

The path to terminating war is paved by urging the intervention of members of the international community. This is always in the minds of ICRC delegates in the field, and, for the unspoken mission, is key. Encouraging third-party intervention is the ICRC's most important task for ending internal

---

* It is possible for the warring parties, if the more powerful side can isolate the confilict, to come to a settlement with little third-party mediation. Russia, with minimal assistance from the OSCE, has moved toward resolving its unfortunate civil war with the Chechens.

wars. Humanitarian tasks also promote this mission, even though they are framed as carrying out the main mission of humanitarian assistance and protection. These five tasks are: 1) protecting and relieving the suffering of civilians; 2) reporting the extent of the suffering and the war crimes that have caused it; 3) visiting prisoners and detainees; 4) reconciling the warring parties and civilians on all sides; and 5) disseminating international humanitarian law (the laws of war) to political authorities, fighters, and civilians.

We will illustrate the ICRC's unspoken mission in action by looking at two cases where third-party intervention was sufficiently massive to bring the warring parties to the brink of peace. The wars in the former Yugoslavia and in Rwanda have moved beyond the indescribable brutality of the past and are now finally able to be resolved with reduced difficulty by the UN or members of the international community. It is true, at the time of this writing, that the United States' policy of weakening the UN and its peacekeeping capabilities by refusing to pay its dues could take away the capability of UN-sponsored efforts to end wars. However, with U.S. troops and prestige committed to fulfilling the Dayton Accord in the former Yugoslavia and with the NATO-led peacekeeping force in place, it is unlikely the war will be allowed to resume. Rwanda is much more problematic.

We will also look at two cases in which political/military third parties are unfortunately not heavily involved: Sudan and Afghanistan. Even so, the ICRC's operations have helped undermine the effectiveness of these wars.

ON THE ROAD TO RESOLUTION: TWO CASES

The Former Yugoslavia

The Context of the War

Both the unity of multinational Yugoslavia and the motives for its recent disintegration go back to World War II. Major elements of the Slovene, Croatian, and Muslim population sided with the German occupation and, with the Germans, proceeded to slaughter about 350,000 Serbs. Major elements of the Serbs and also portions of the other nationalities conducted guerrilla warfare against the German side under the communist leadership of Josip Broz Tito. With Allied help, Tito won. He proceeded to unify the state under a federation of republics and autonomous areas dominated by the Communist Party. In addition to the stature of Tito and the brutal efficiency of the Communist police state, Yugoslavia was held together by the threat of invasion from the Soviet bloc, still a possibility after Stalin's 1948 failed attempt to subvert Tito.

One by one, the three forces holding the federation together disappeared or declined. Tito died in 1981. Communism virtually collapsed in Europe in 1989, and the Soviet Union fractured and vanished in 1991.

National hatreds flared in Yugoslavia. Religion—the Slovenes and Croats are Roman Catholic, the Bosnians Muslim, and the Serbs Orthodox Christians—reinforced the national enmity generated in World War II.

Independence movements, first in Slovenia, then in Croatia, Bosnia-Herzegovina, and Macedonia, arose under leaders who used extreme nationalism to gain and keep power. No provisions were made for the rights of the minority Serbs who made up 12 percent of Croatia and 32 percent of Bosnia. In 1991, resisting dismemberment, Serb-dominated Yugoslavia moved its armed forces against Slovenia and Croatia. It only succeeded in occupying the Serbian-populated areas of Croatia with the help of local Serb militias. However, keeping the former Yugoslavia totally intact proved beyond the power of Belgrade.

The UN Security Council then authorized and dispatched a protection force (UNPROFOR) to stabilize the de facto division of Croatia. In April 1992, after Bosnia proclaimed independence, Serb forces moved against the Muslim-dominated government of Bosnia. UNPROFOR's role expanded to Bosnia and focused on protecting humanitarian relief efforts. The Security Council imposed an economic embargo on Belgrade and an arms embargo on all the warring parties. With the urging of the German government, the West recognized the breakaway republics in May 1992. This had the effect of lining up the West against Serbian Yugoslavia.

The Serbs in Croatia and Bosnia wanted to remain connected to Belgrade in a "Greater Serbia." Their strategy was to remove all non-Serbs from areas they could dominate, then hold that territory and await partition. Thus began their campaign of ethnic cleansing using combat, rape, murder, and arson. The Serb campaign's heavy dose of terrorism was designed to drive out Muslim and Croat civilians.

The Croats' goal was to regain their land occupied by the Serbs in Krajina and Eastern Slavonia, and to link up with the Croats in Bosnia. Their strategy was to build international support for an expanded UN role in Bosnia, re-arm, create an occasional tactical alliance with the Bosnians, and wait until the Serbs were weakened before beginning their military offensive against them.

The Muslims wanted to keep Bosnia whole and themselves in charge. Their strategy wedded them to U.S. diplomacy, UNPROFOR protection, and smuggled munitions. They knew that maximizing outside political, military, and economic support was essential since they were the weakest of the three warring parties.

The Process of Ending the Wars

The Dayton Accord of November 21, 1995, ended the Croatian and Bosnian wars. They had produced 150,000 or so deaths, 2 million refugees and displaced persons, and unspeakable suffering. The Accord was the product of a series of forces that made war dysfunctional. Specifically: 1) massive third-party intervention prevented victory by any side; 2) humanitarian relief sustained, and to some degree protected, the civilian population; 3) exhaustion and frustration made the warring parties and the intervening forces ready to settle; and 4) a plan was packaged that matched the situation on the ground and gave each of the warring parties something more than they could get by continuing the war. Each factor will be briefly reviewed, and then the ICRC's contribution will be described.

1. Massive third-party intervention prevented victory. It did not start out massive. UNPROFOR was understaffed, weakly armed, and scattered over Croatia and Bosnia. Negotiators from the European Union, NATO, and finally the Contact Group of France, Great Britain, Russia, and the United States, therefore, lacked the military backing to insist on a settlement. In addition, any plans that called for the partition of Croatia or Bosnia were opposed by the United States because it was committed to preserving the two states. Naturally, the governments of Croatia and Bosnia opposed being partitioned by the Serbs. Any plans that provided for the unity of Croatia and Bosnia were opposed by the Serbs. The war dragged on. The UN Security Council, bold in passing resolutions and weak in carrying them out, voted to designate Srebrenica in Bosnia a "safe area" in April 1993. In June, five other safe areas—Tuzla, Sarajevo, Garažde, Zepa, and Bihać—were designated. Serb attacks on the safe areas brought weak, ineffective NATO air strikes. In response, Serb forces took British and French units of UNPROFOR hostage. French President Chirac demanded action. When Serb forces overran the safe areas of Zepa and Srebrenica in July 1995 and then murdered fighting-aged males by the thousands, the Security Council, referring to Chapter VII, authorized "all necessary means" to stop the slaughter. UNPROFOR forces regrouped behind Bosnian lines. A NATO rapid reaction force was deployed. A month after a particularly devastating mortar attack on a Sarajevo market in August 1995, this force began shelling Serb positions. At the same time, NATO launched heavy air strikes on Serb communication, command, and military facilities.

Seeing the Serbs weakened, the Croat forces struck at the Serb positions in Krajina on August 5, 1995, pushing both Serb fighters and civilians, almost a quarter of a million people, out of Croatia into Serb-held areas in Bosnia. The Bosnian Muslim forces, in cooperation with Croatian forces in Bosnia, then went on the offensive. The Serbs then suffered ethnic cleansing. Surprisingly to some, Yugoslav President Slobodan Milošević failed to reinforce the Serb militias in Croatia and Bosnia. He failed to do so in order to create the conditions for a settlement. He was satisfied with the half of Bosnia he could get in a deal, wanted to discredit Bosnian Serb leader Radovan Karadžić (and thus remove a rival), sought to end the crippling trade embargo on his country, and desired to restore normal relations with the international community.

Once the situation on the ground looked like a map that could be incorporated into a settlement, NATO forces commanded that the Croats and Bosnians halt their offensives.

2. Humanitarian relief sustained, and to some degree protected, the civilian population. The over 2 million refugees and displaced people elevated the role of the UN High Commissioner for Refugees (UNHCR). The UNHCR delivered, with UNPROFOR escort, massive amounts of food to the protected areas. While two of the six protected areas eventually fell, starvation was not the cause. And until Zepa and Srebrenica were overrun, the record of UNPROFOR in protecting civilians was somewhat effective. Certainly, without UNPROFOR and the 300 NGOs in the field, Serb forces, with the backing of Belgrade, would have been able to seize all of Bosnia-Herzegovina. When the Serb forces went on the offensive in the summer of 1995, it only brought increased political, military, and humanitarian intervention, and this led to exhaustion and frustration.

3. Exhaustion and frustration made the warring parties and the intervening forces ready to settle. Coinciding with the fall of Zepa and Srebrenica, the U.S. Congress threatened to lift the arms embargo on Bosnia. To counteract this move, which would have revealed President Clinton's loss of foreign-policy control, and to satisfy the blunt urging of President Chirac, Clinton moved to a policy of peace enforcement. The rapid-reaction force and NATO airpower paralyzed the Serb forces and isolated them from Belgrade. The Bosnian-Croat offensive then redrew the map. It was clear to all sides that further military efforts would not be fruitful. NATO forces would not allow it. Dayton became the best option.

4. A plan was packaged that matched the situation on the ground and gave each warring party something more than they would get by continuing the war. Representative of Bosnia, Croatia, and Yugoslavia (speaking for all Serbs) were sequestered at Dayton's Wright-Patterson Air Force Base in late October 1995. Tough-talking Assistant Secretary of State Richard C. Holbrooke hammered out the Accord. Zones of separation would be created and patrolled by IFOR troops under the control of the North Atlantic Council. Bosnia would remain a state with a federal government in Sarajevo; it would consist of two political entities—a Muslim-Croat Federation and a Serb Republic—each with its own government and army. This unusual configuration had some logic behind it. It was acceptable to the three parties and had the potential to evolve from a de facto two-state situation to an eventual single federal state, similar to Switzerland or Canada.

POWs would be exchanged. The UNHCR would manage the repatriation of refugees. The Organization for Security and Cooperation in Europe (OSCE) would set up an election commission. And Carl Bildt, the former prime minister of Sweden, would act as the senior civilian diplomat to mediate disputes between the parties. All parties were obliged to hand over indicted war criminals to the War Crimes Tribunal at the Hague.

The Role of the ICRC in the Resolution of the War

The ICRC, in Hollywood talk, would win the Oscar for best supporting actor. Its 1994 *Annual Report* defined its role.

> In the ICRC's view, peace-making efforts could only succeed if backed by the international community through a coordinated approach at the political level. Until this was achieved, the institution felt it was the duty of governments to prevent any escalation of the conflict and to lend their full support to emergency programs.[1]

At every turn, the ICRC urged intervention to dampen the war. Whether the carnage was the shelling of the Sarajevo market or attacks on the Goražde protected areas or murders in the Bosnian town of Prijedor, the "ICRC made constant efforts to draw the attention of the international community to their [the civilians] plight."[2] It called a meeting in Geneva in September 1994, inviting all Geneva-based diplomats and reminding them, in the words of President Sommaruga, "of their collective obligation to ensure that the provisions of humanitarian law were respected in all circumstances."[3] Furthermore, the ICRC lobbied the Contact Group, the President of the Security Council, and the Secretary-General, complaining that its reports on the humanitarian implications of ethnic cleansing had been ignored.

The area in which the ICRC had the most effect in making the war riper for resolution was in its humanitarian relief that sustained and to some degree protected the civilian population (our second factor discussed above). This was its "highest priority."[4] By helping to separate civilians from the warring parties, it contributed to the latter's exhaustion and frustration and thus to their disposition to settle the conflict. All five tasks listed on page 68 played a role.

First, considerable resources were devoted to sustaining and protecting civilians, whether Croats, Serbs, or Muslims. ICRC delegates—270 since the start of the war and backed by 1,200 local staff—maintained 30 delegations, sub-delegations, and offices throughout the former Yugoslavia. They coordinated the activities of teams of workers from 17 National Societies. They annually spent over $100 million, visited an average of 9,000 detainees, distributed more than 100,000 tons of relief supplies to over a million civilians on all sides, and operated 230 medical facilities.[5] The ICRC was often the only humanitarian organization operating in the most active war zones.

In May 1992, the ICRC brought representatives of all the warring parties together *for the first time* and got them to sign four agreements in the following months that stipulated that prisoners and detainees would be treated humanely and that civilians "be protected against dangers arising from military operations."[6] Whenever ICRC delegates learned of violations of these agreements, they would inform the warring parties, first locally and then at the top, that violations must be stopped. If this failed to remedy the violations, ICRC leadership in Geneva would then inform all the states party to the Geneva Conventions "to remind them of their obligations."[7] In June 1995, ICRC President Cornelio Sommaruga appealed to the warring parties to honor their humanitarian commitments made in 1992. This message was repeated in August 1995, when ICRC President Cornelio Sommaruga visited the authorities in Belgrade, Pale, Sarajevo, and Zagreb. When witnessed, descriptions of violations of international humanitarian law were passed on to the warring parties for remedial action, and, if not forthcoming, were passed on to the international community.[8]

Nik Gowing, Diplomatic Editor for ITN's Channel Four News in London, reporting on his study of the effects or real-time TV on government policy, described as successful the ICRC's "Sherlock Holmes" efforts to pass on to ITN the horrors of Serbian POW camps.[9] The governments that had peacekeepers in Bosnia then put pressure on the offending parties. While not putting an end to ethnic cleansing, the ICRC believed that its "measures have undoubtedly save thousands of lives by helping prevent massacres such as those that took place during the early stages of the war."[10]

> On-the-spot presence and direct contact with the victims enable the ICRC to monitor the situation in an impartial manner and report on it independently and confidentially to the highest authorities. Thus its presence has a deterrent effect; this is one of the reasons why it is often so difficult for the ICRC to reach the victims, and why access remains its top priority.[11]

For example, it took ICRC delegates several weeks to get access to detainees in Mostar in 1993, and the same for those around Bihać. Stating the obvious, the ICRC concluded that "the parties did not want witnesses to their unacceptable practices, including the execution of prisoners."[12] Delegates would urge the warring parties to release civilian detainees, "unconditionally and unilaterally."

Exchanges involving equal numbers only caused the side with fewer detainees to seize more. To encourage their release, delegates would strive to register every detainee. In this way, they could check during future visits whether any of them were missing and thus prevent their disappearance. The lists were evidence that could be used in the UN War Crimes Tribunal, although ICRC delegates never agree to testify at trials in order to maintain their neutrality. The strategy for detainees had some effect, although mass executions continued. In late 1992, for example, some 5,600 detainees were released.[13]

It was the ICRC in late 1992 that suggested to UNPROFOR that protected areas be established to provide safety to minority populations.[14] Six were established the next year. At the same time, assistance—such as providing water supply systems—was rendered in Serb civilian areas. The principles of neutrality, impartiality, and independence were maintained.

Working directly with the warring parties, the ICRC conducted dissemination sessions to make them aware of the rules of war as established in the four Geneva Conventions and Additional Protocols. It put the same message into leaflets and on 80 radio stations and 20 TV stations, and in 25 newspapers. By 1994, these efforts "were reaching most of the military units in Bosnia-Herzegovinia. The ICRC also gave talks on international humanitarian law for UN contingents."[15]

Protecting civilians in the former Yugoslavia was the ICRC's greatest challenge and achievement. It also presented ICRC policymakers with a major dilemma. In my interviews with officials, more than one raised the dilemma without my asking. One ICRC publication put the matter succinctly.

> Ethnic identity has become central to many conflicts, and the phenomenon of "ethnic cleansing" more widespread. Humanitarian organizations are faced with the desperate task of trying to protect people whose death or removal is the objective of the war in the first place and whose lives are worth nothing to their oppressors. In such cases, the delegates are faced with a gruesome Hobson's choice: between protecting victims where they are, or helping them to flee to relative safety. Either way, both the victims and the Red Cross Principles are somewhat compromised. If the ICRC helps the victims stay put, they may become targets and the ICRC may be accused—as it was in the former Yugoslavia—of putting them in further danger. If they leave, the victims lose their homes and the ICRC may be accused of facilitating ethnic cleansing.[16]

In most cases, the ICRC accepted, as a last resort, the lesser evil and helped civilians gather and, under ICRC escort, move to safer areas. Better to save lives than uphold principles that would sacrifice lives. And the effect was

An Escort for Displaced Persons in the Former Yugoslavia.

After the 1995 Croatian offensive in its Krajina region against the Serbs, over 200,000 Serbs fled to their stronghold around Banja Luka in Bosnia. The ICRC provided the escort. By doing so they protected the refugees as they moved out of harms way and facilitated the regrouping of populations to single-nationality areas. This helped transform the war into a defensive one, whereby it could stalemate and thus become ripe for negotiations at Dayton. *(Photo courtesy of ICRC, by Thierry Gassmann)* to re-group the three nationalities to areas controlled by their own authorities. This, as it proved, facilitated the Dayton Accord by making the zones of separation easier to draw.

The wars in the former Yugoslavia raged over four years. They did not produce clear-cut victories. The Dayton Accord gave each side something—perhaps the Croats and Serbs more than the Muslims. By assisting and protecting civilians, the ICRC dissipated the wars' fuel. The chances for more peaceful relations improved, and the ICRC and other NGOs played a major role in making that option the most desirable for all the parties.

The Dayton Accord did not end ICRC involvement. During the negotiations, the ICRC worked behind the scenes, reminding the parties of the importance of the plight of displaced populations and ethnic minorities and of the need for the unconditional release of detainees. As a result, the ICRC was charged by the Dayton Accord with establishing The Central Commission for Prisoners and Persons Unaccounted for; this required tracing persons and supervising the release and transfer of prisoners.[17] In addition, the ICRC and the International Federation worked with the World Bank and others to plan for "peace building" (the rehabilitation of Bosnia).[18]

A far worse case, with more meager results (although the killing is now a mere fraction of that in the past), is the ICRC's effort in Rwanda.

The Rwandan Civil War

The Context of the War

From the sixteenth century until 1890, the area that is now Rwanda and Burundi was ruled by a Tutsi monarchy, even though the Hutu tribe made up over 85 percent of the population. Germany incorporated the region into German East Africa in 1890 but kept the monarchy as a front. The Belgians acted similarly when they took over the colonies in 1919, after Germany's defeat in World War I. In 1959, as independence approached, the Hutus rose up and overthrew the monarchy. Many Tutsis fled to neighboring territories, especially to Uganda in the north. A UN-supervised election in 1961 brought a Hutu-dominated republican party to power, and Rwanda's independence was recognized the following year.

Coups and periodic violence between Hutus and Tutsis plagued the new state, which received outside military assistance from Belgium and France. Fighting intensified in 1990 when Tutsis backed by Uganda attacked from the north. They fought under the name of the Rwandan Patriotic Front (RPF). A cease-fire, optimistically called the Arusha Peace Agreement, was brokered under the auspices of the UN in 1993, and 2,500 peacekeepers organized as the United

Nations Mission in Rwanda (UNAMIR) were deployed. In early 1994, radical elements in the Rwandan government began planning a "final solution" for moderate Hutus and Tutsis. On April 6, 1994, the plane carrying Hutu President Juvenal Habyarimana was shot down as it was landing at Kigali airport. Tutsis were blamed by the radical Hutu leaders. Inflamed by Hutu calls for retribution on the radio, Hutu militias began the massacre of moderate Hutus and Tutsis.

UN commander Romeo Dallaire of Canada called for an additional 2,500 troops to stop the killing, but the Security Council chose not to expand the mandate. Ten unarmed Belgian peacekeepers were murdered along with the prime minister, an incident that led to UNAMIR's regrouping and, by Security Council order, virtual withdrawal until only 270 peacekeepers were in place. This callous response by the international community facilitated the genocide. Some estimates put the number of deaths, largely by machete chops, at over half a million. Few Tutsis outside of the RPF-controlled areas escaped.

Two days after the massacres began, the RPF went on the offensive. By June 1994, the Hutus, including the militias that organized and conducted the killings, were in flight. Over 2 million refugees poured into Tanzania and then Zaire. Cholera, killing 50,000, swept the camps in Goma, Zaire, that had only polluted water to drink. Belatedly, the Security Council expanded the mandate in May, but few members were willing to commit troops. Contingents of French troops arrived to provide some protection to the Hutus in the southwest of Rwanda.

The Process of Ending the War

As with the former Yugoslavia, the war fizzled out when the populations regrouped. The RPF took control of Kigali in July and secured the rest of Rwanda soon after. There were occasional Tutsi reprisals against Hutus still in Rwanda, but not many. UNAMIR's 1,800 troops provided some security around Kigali, but by mutual agreement these forces were withdrawn in March 1996. In the camps, especially those around Goma, Zaire, a flood of TV crews from Europe and the United States recorded the plight of the Hutu refugees. This was soon followed by a flood of military technicians who constructed water purification and sanitation installations. Hundreds of NGOs, their activities coordinated by CARE, joined the UNHCR in providing food, medical assistance, and other relief, knowing full well that some of this aid was going to the very military and militia members that committed the genocide against the Tutsis.

The war froze. Efforts by the UNHCR and the Zairian military to repatriate the refugees proved ineffective, as fear and Hutu militia leaders (who needed a civilian base of support for their plans to invade Rwanda) kept them in the camps. More babies were born in the camps in the two years after the genocide and RPF victory than the number of people who were repatriated. At least the massive dying stopped.

Burundi, with the same volatile tribal mix of Hutus and Tutsis, suffered occasional bloodlettings.

Where was the plan to put Rwanda back together? Where was the international leadership to carry it out? In effect, the borders of the neighboring states separated the warring parties and so became the "peacekeepers." But not very good ones. Cross-border raids into Rwanda by Hutu militias increased in 1996. This incited the Tutsi-led Rwandan government to arm Zairian rebels, who then attacked the Hutu militias and dismantled the refugee camps near the border. As the Hutu militias fled deeper into Zaire, hundreds of thousands of Hutus, freed from militia control, streamed back to their villages in Rwanda. By early December 1996, less than six weeks after the Zairian rebels attacked the camps, a half million Hutus found a largely peaceful homecoming. More refugees soon followed, ejected from Tanzania by Tanzanian troops.

Whether in camps or on the road, the ICRC, the Federation of Red Cross and Red Crescent Societies, the UNHCR, the World Food Program (WFP), and other organizations worked to sustain the lives of civilians and to create conditions for reconciliation.

The Role of the ICRC in the Stabilizing of the War

There are limits to what the ICRC can do to stop wars and encourage a peace settlement. In Rwanda, it was not for want of trying.

Even before the assassination of the president, the ICRC was providing food to over 600,000 displaced persons in Rwanda.[19] As soon as the massacres began, ICRC delegates in Kigali, via Geneva, informed their UN observer mission in New York of the desperate need for beefing up UNAMIR to protect civilians. The ICRC, according to delegate Daniel Augstburger, "was screaming at the Security Council for its failure to move."[20] The first task of getting massive third-party intervention in the war was never accomplished. Initially, the ICRC was largely on its own. In 1994, it mounted its biggest world-wide relief operation in and around Rwanda. In the beginning, the courage shown by delegates and local staff, alone and vulnerable before bloodthirsty militias, drew widespread praise in the world of humanitarian relief. The ICRC judged the effort worth the cost.

> The horrific turn of events in Rwanda in April forced practically all foreigners to flee the country, while the United Nations withdrew most of its troops. The result would almost certainly have been more catastrophic still if the ICRC too had left. As it was, the delegation in Kigali was able to keep the outside world informed of the situation by

Distributing Food in Rwanda.

The ICRC sustains the victims of war, demonstrating that warlords cannot deliver even basic staples to their people. This relief weakens the civilian base of support for the warring parties, even at the cost of creating a civilian dependence upon the largesse of humanitarian agencies. For the long term, the ICRC establishes programs to help civilians feed themselves. *(Photo courtesy of ICRC, by Thierry Gassmann)*

> satellite telephone. This communications link was the only neutral source of information in a contest rife with lies and rumour, and as such was absolutely vital. It also acted partly as a stabilizing factor, while the presence of ICRC delegates among groups of civilians at risk undoubtedly saved many thousands of lives. Although the massacres subsided, right up until the end of the year the spectre of renewed bloodshed loomed over the whole of the Great Lakes region. The situation remained extremely unstable, as over two million refugees remained camped close to the Rwandan borders, among them troops of the former government, which were exiled in Zaire.[21]

The failure to get peacekeepers involved in protecting civilians put the burden on the ICRC and eventually on other NGOs and UN humanitarian agencies. The initial focus was within Rwanda. Emergency surgical care was provided to the wounded in Kigali. In the first week, 25 tons of medical supplies were delivered, mainly to Kigali Central Hospital.[22] Médecins sans Frontières worked with the ICRC under the Red Cross banner. Other hospitals were created as needed, often following the bloody trail of the Hutu militias. Many medical units came under fire. Pockets of potential victims, about 50,000 around the capital, were visited by ICRC delegates and food brought to them daily. By June, the ICRC was feeding a half million displaced people within Rwanda, and this grew to 1.2 million by September. Four aircraft and over 130 trucks were used to deliver 89,000 tons of relief supplies.[23]

Since contaminated water is one of the prime carriers of disease, the ICRC had teams restoring sanitary water in nine of the country's major cities. An indication of the scope of this activity is the tonnage of chemicals delivered: 300 of aluminum sulfate, 50 of chlorine, and 300 of lime.[24]

As the RPF advanced and took prisoners, the ICRC found itself increasingly involved in prison work—restoring water supplies, registering 16,000 detainees, and providing blankets, soap, and utensils. In 1995, the ICRC actually participated in the construction of a prison, "judging it to be the only way to contribute to the survival of the detainees and show an example of what the authorities could do to improve conditions."[25] By August 1996, over 83,000 detainees were visited and registered.[26]

It is difficult to judge the effects of dissemination efforts since international humanitarian law was so widely abused. Yet, the ICRC did not stop trying. And even after the RPF victory, its military received training and literature on how to treat civilians if fighting resumed.

Well before the 2 million or so refugees streamed out of Rwanda, the ICRC found itself in a good position to help. In Goma, where the ICRC had been operating since 1993 in connection with tribal conflicts, a logistics base

An ICRC Medical Unit in Rwanda.

A camp of displaced persons is tightly clustered near the ICRC clinic, giving the residents some measure of protection. The business of the camp is not war and it cannot legally be used for war, even though ex-warriors may remain there. It must frustrate those who want to fight to live in a village where its energy, resources, and personnel are not geared to support the war. In time, displaced persons may become accustomed to peace. *(Photo courtesy of ICRC, by Patrick Fuller)* with 1,000 tons of food was in place. Relief, therefore, was immediate.[27] After extensive media coverage and a corresponding international relief effort, the ICRC in August 1994 turned over its operations in the camps to the National Societies and their International Federation.

In 1994, the ICRC spent about $130 million bringing some measure of humanity to the genocide.[28] It had done all it could to bring in the UN peacekeepers. It had saved tens of thousands of civilians. It had helped stabilize the aftermath of an uncivil civil war, all the while showing a humanity worth emulating by the butchers. Even with some measure of stability, three immediate issues remained on the ICRC's agenda. First, there was insufficient rehabilitation by the international community. In 1995-96, the ICRC provided emergency food aid, seeds, tools, and medical care to about 400,000 residents and displaced persons every month.[29] Second, the Rwandan criminal justice system was in shambles. Of the 45,517 persons in 14 prisons during this period, the ICRC met all their food needs and made sure all had adequate medical assistance and water.[30]

When the Zairian rebels attacked the refugee camps in October 1996 and defeated the Hutu militias that controlled them, a half million Hutu civilians fled the camps. The advice given by virtually all the humanitarian agencies to the refugees was clear. Go home. In addition, the Zairian rebels also wanted an end to the camps, and forcibly closed the Red Cross facilities there. This kept the Hutu refugees on the road, but without the means to ensure their survival. The ICRC quickly established mobile aid stations along the roads and dispensed high-protein biscuits and treated refugees suffering from de-hydration. Inevitably, children became separated from their parents. The ICRC, with typical speed, began a feeding program and tracing activities for the children. When traced, the children were transported back to their families and villages by the ICRC.[31]

The ICRC lost the battle against the war, but in the long run, the war in Rwanda further discredited war's reputation for resolving conflict at reasonable costs. The Tutsis lost half their members, and as of early 1997 about one-tenth of the Hutu population still remained outside of Rwanda.

ON THE ROAD TO DYSFUNCTION: TWO CASES

The wars in Sudan and Afghanistan are two cases in which humanitarian assistance is not accompanied by third-party political mediation or peacekeeping. This situation, as mentioned before, creates a dilemma for relief agencies. To provide food, medicine, and shelter to civilians who are under the domain of warlords is to add fuel to the war. The warlords want to control and distribute relief supplies in order to keep their civilian base of support for the war effort. The struggle for NGOs, therefore, is to keep relief separate from the war, but knowing that the struggle will not always be won. To win it, relief efforts would have to be halted in most cases, but to do that would inevitably result in civilian suffering. So, to ease civilian suffering *and* to prevent any side from being wiped out (thus making the war dysfunctional—the unspoken mission), the ICRC and other humanitarian agencies provide relief.

The War in Sudan

The Context of the War

Sudan is a country with no unifying principle, and the attempt to impose one has divided the country north and south. The Ottomans ruled Sudan from Egypt until the revolt led by the Mahdi in 1885. Then the British took over. Egypt shared a culture with the north of Sudan shaped by the Nile, Arabic, and Islam. The southern population (now 4 of the 26 million Sudanese) had a different culture—black African, with local vernacular languages and Christian or animist religions. Until 1860 and even beyond, the north used the south as a source of slaves, a history that the people of the south still resent. The British sustained their hegemony in Egypt and Sudan at the turn of the century through cultural divisions, ruling the three southern provinces separately.

Independence in 1956 brought a southern revolt. This faded somewhat in 1969 after a relaxed government was installed by a coup in Khartoum led by Gaafar al-Nimeiri. After fourteen years in power and losing popular support, Nimeiri opted to rebuild his political base via Islam. He commanded that the *sharia* (Islamic law) be the law of the land. This was a direct cultural threat to the south and it re-ignited the southern insurgency.

Nimeiri was replaced by Sadiq al-Mahdi, who was replaced in a coup in 1989 by Colonel (soon to be General, of course) Omar Ahmed al-Bashir. The real power in Khartoum, however, is the head of the National Islamic Front (NIF), Hassan al-Turabi. Turabi, who holds a doctorate from the Sorbonne, uses the NIF to infiltrate the leadership structures in the north—particularly the government bureaucracy—in order to build an Islamic state. He does not hesitate to use a full range of police controls and rigged elections to stay in power. He is backed internationally, it is alleged, by Islamic political factions in Saudi Arabia and Iran.

In the south, the insurgents formed the Sudan People's Liberation Army (SPLA) under John Garang. He is a Dinka and holds his Ph.D. from Iowa State University. Originally backed by the Marxist regime in Ethiopia, the SPLA split when that backing ended with the change of regime in Addis Ababa in 1991. Riek Machar, a Nuer with a Ph.D. from the British University of Bradford, formed a rival South Sudan Independence Army/Movement (SSIA/M). Most of the killing since 1991—the death toll is well over 1 million since 1983—has resulted from fighting between the two southern warring parties. Garang seems to be prevailing, and his SPLA is backed by governments that oppose the spread of Islam by terror, principally Egypt and the United States.

The three warlords are giving the Ph.D. a bad name. All are ruthless, employ murder and starvation, and burn the villages of their opponents. All plunder relief operations. Relief workers privately condemn them in the most unflattering terms as they witness death, starvation, and, what is most disappointing, a blasé neglect of the war by the international community. One journalist, Nik Gowing, quotes a U.S. diplomat who described Sudan as "Somalia without CNN."[32] It is one measure of a war's marginal effects on the international community's political and economic interests when no third party attempts to mediate the war. It is an absolute measure of the war's inconsequential nature when, with hundreds of thousands of deaths, CNN fails even to cover the conflict.

The ICRC's Efforts to Make the War Dysfunctional

The ICRC made some initial attempts to get the Security Council to approve UN mediation to end the war in Sudan. It proved futile. Better than nothing, the UN since 1989 has been managing Operation Lifeline Sudan out of Lokichokio in Kenya. Every week $2 million in food aid is delivered to the south by Operation Lifeline.[33] The ICRC at one time contributed $1 million a month in food relief, but now it concentrates on supplying seeds, tools, fishhooks (521,000 of them in 1994 and even more in 1995), twine, and mosquito netting (350,000 meters worth in 1994 and slightly less in 1995).[34] Agricultural rehabilitation is a prime concern of the ICRC delegates in order to enhance civilian self-sufficiency. This activity is neglected by other relief organizations.

Another concern is medical. Patients from the south are brought to the ICRC hospitals in Juba (near the Ugandan border) and Lokichokio. Over 7,500 surgical operations were performed and the ICRC supplied 100 tons of medical supplies to 70 health-care centers in 1995.

The overall task of the ICRC is to "enhance the self-sufficiency of the population in southern Sudan."[35] As long as the civilians can be maintained, two antiwar tasks of the unspoken mission can be accomplished. Civilian self-sufficiency weakens the SPLA and the SSIA/M by cutting reliance on these unsavory warriors, but the leakage of relief aid also sustains these warriors sufficiently to prevent a northern victory.

The ICRC also cooperates with the Sudanese Red Crescent Society in Khartoum in providing services to the hundreds of thousands of displaced people who have fled the war in the south. Constructing clean water facilities has been one focus in the north. This impartiality is helpful in negotiations with the al-Bashir government to receive authorization to deliver relief supplies to the south, and, so far unsuccessfully, to visit detainees.

As with a similar, seemingly never-ending war in Afghanistan, the ICRC does what it can to relieve the suffering caused by war and to prevent its success as a strategy. It is one thing to make war dysfunctional; it is another to induce powerful political agents and peacekeepers to shape the end game and engineer a settlement.

The War in Afghanistan

The Context of the War

In the nineteenth century, Britain and Russia played what came to be known as the "Great Game" in the lands between the czarist empire and the British Raj of India. The Russians pushed south, attempting to enroll Persia and Afghanistan in their sphere of influence. The British in India aimed to stop them by keeping these borderlands acting at least as neutral buffer zones if not under outright British control. Decades of intrigues, treaties, espionage, and armed expeditions followed. Finally, in 1907, Britain and Russia agreed to allow Afghanistan to remain neutral. It remained so until the Communists staged a successful coup in 1978.

A modernizing Marxist "revolution from above" was begun by Nur Mohammed Taraki. Commissars from Kabul forced agrarian reform on the countryside and stripped tribal leaders and Muslim clergy of their authority. Resistance to these measures brought executions. Executions brought the formation of *mujahedin* (holy warriors) and a civil war in late 1978.

The war did not go well for Taraki or his successor, Hafizullah Amin. The United States, following a policy designed by national security advisor Zbigniew Brzezinski, intervened in the war by supplying arms to the *mujahedin.* Amin then indicated that he was considering changing Cold War sides. To prevent this, to contain the spread of political Islam into Soviet territory, and to counter U.S. intervention, the Soviet Union invaded Afghanistan in December 1979, killed Amin and his family, and began a nine-year occupation.

Supported by the United States, China, and Iran, the *mujahedin*—a loose union of tribes—fought the 120,000 Soviet troops to a standstill. The war increased in cost to the Soviets in 1986 when the United States provided the insurgents with shoulder-fired stinger anti-aircraft missiles. Soviet helicopters, key weapons in moving and protecting troops, proved highly vulnerable to the stingers. These costs reinforced Gorbachev's conclusion that the revitalization of the Soviet Union required the end of the Cold War. Following an agreement brokered by the UN, Soviet troops began withdrawing in 1989.

For three-and-a-half years, the Marxist government held out against the *mujahedin,* finally falling in April 1992. By this time, with the Cold War as history, the war had faded as a foreign-policy concern to all but Afghanistan's neighbors. The immediate concern of Pakistan and Iran was the 4 million refugees encamped within their borders (out of a total 15.5 million population in the last census before the communist takeover).[36] Both states wanted the war over.

Unfortunately, they wanted different *mujahedin* factions to win. Immediately after the fall of the government the Soviets had left behind, the holy warriors became less whole. One faction, backed by the northern Tajik forces of Ahmad Shah Massoud, took the presidency under Burhanuddin Rabanni. Lesser posts went to Gulbuddin Hekmatyar's Hezb-i-Islami, a largely Pushtun militia from the south. The falling-out began that same year (1992), and found Hezb-i-Islami's forces bombarding Kabul. This new civil war stalemated. Seeing an opportunity, a new military force, named Taliban after its student roots among Pushtuns in the south, began an offensive against all opposition in 1994. It, too, besieged Kabul, adding to the 40,000 or so people killed in Kabul since 1992. Taliban took control of the war-torn capital in September 1996, and imposed strict Islamic law. The Pakistani government backs the Taliban, hoping by doing so to rid itself of refugees and win the support of its own Pushtun (called Pathan in Pakistan) ethnic population. This move is opposed by Iran, Russia, and India, who provide arms, ammunition, and supplies to a coalition of anti-Taliban forces based on Tajik, Uzbek, and Hazaras ethnic groups. These groups control the northern one-third of the country.[37] In early 1997, the war reached another dismal stalemate.

One million deaths. Taliban proving to be as bloodthirsty as their older *mujahedin* opponents. Kabul largely a wasteland. The economy primitive at best. Ten million land mines still on duty producing 200 casualties a day, according to relief workers.[38] Above all, neglect. The United States, which fed an estimated $5 billion in arms and supplies to the *mujahedin,* stopped all its direct relief activities in 1994. UN humanitarian assistance has fallen, leaving much of the burden to the ICRC and a handful of NGOs.

The ICRC's Efforts to Make the War Dysfunctional

The ICRC has promoted the efforts of the UN Special Envoy and those of a representative of the Organization of the Islamic Conference. These efforts, unfortunately, "achieved no tangible results."[39]

The 13 ICRC installations in Afghanistan, as they do in Sudan and Somalia, serve to protect civilians from the ravages of war and wean them away from the warring parties by making them more self-sufficient. When the siege of Kabul in early 1996 severely reduced food supplies in the capital, the ICRC organized an emergency airlift from Pakistan to an air base northeast of Kabul still under government control. Several flights a day eventually brought in

1,000 metric tons of wheat. As reported in *The New York Times,* "The airlift planners hope to show the besieging guerrillas that they cannot starve out the city."[40] Sieges are a war strategy designed to starve out both soldiers and civilians. The ICRC and other relief organizations countered that strategy to the point that Taliban forces had no alternative but to make a lightening and relatively bloodless assault on Kabul. Ironically, this probably saved lives because the siege and bombardment killed civilians daily.

In this dry land, underground irrigation channels are essential for watering the orchards and fields. Repairing them in cooperation with local farmers is the work of ICRC agricultural teams. According to one agro-team leader, Mr. A. R. Noori, "we want the local population to take an active part and, if possible, to be self-sufficient even during wartime."[41] Medical assistance is provided to 65 facilities on all sides of the war. Of special need, in light of the horrendous dismemberment caused by land mines, is the production and distribution of orthopedic devices. In 1995, 2,216 patients were fitted with artificial limbs.[42]

A new set of problems confronted the ICRC after the takeover of Kabul by Taliban. Taliban banned women from working. This left about 25,000 women, mostly widows. at home with their children and with no one to support them. Thomas Gartner, an ICRC delegate in Kabul, called their situation "desperate" since many women had seven or eight children to feed. Fortunately, the ICRC had cultivated good relations with Taliban—as it tries to do with all warring parties—and this facilitated a program through which the ICRC provides food supplements to 15,000 women and their families.[43]

By providing food, seed, tools, technical assistance (such as for irrigation and drinking water supplies), medical care, the tracing of family members, neutral intermediaries for the exchange of the dead, and the registration of detainees, the ICRC is substituting services normally provided by political authorities, which in this case are also the warring parties. These ICRC services, in conjunction with the services of the Afghan Red Crescent Society, have no political strings attached and thus are attractive to civilians. Because of this, they have an unusual and important effect on civilian attitudes towards the war.

Compared to the benefits received through humanitarian assistance, the actions of the warring parties in producing insecurity, unemployment, famine, maiming, death, and disease become highly distasteful. This translates into strong antiwar attitudes and fierce disdain for political leaders who pursue the war. The popular base for pursuing the war erodes. War weariness soars.

Non-political humanitarian work is subversive to war. It is subtle sabotage. In every case mentioned in this study, hostile attitudes towards war have been engendered by ICRC work.

Among mothers who have lost husbands and children.

Among women who have no means of support.

Among the elderly who see their villages disappear.

Among children who are denied freedom to play.

Among many young males who are conscripted by the warlords, and who rarely see progress towards ending the war.

If people pay close attention to media accounts of current wars, they quickly discover that much of the content of these reports deals with civilian suffering and their anger about the war itself. Rarely are there cheers. Parades are not performed because few would wave the flags.

What follows are some typical newspaper accounts from *The New York Times* and *The Christian Science Monitor* of antiwar feelings of those served by humanitarian agencies in Afghanistan:

> "At the hospital, the children spend their days staring at the roof and walls scarred by bullet holes, vivid reminders of this country's 16-year-long conflict."[44]
>
> "Among the most visible signs of the destruction facing this country are the almost-empty playing fields around schools."[45]
>
> "So dispirited is the mood that it is common to hear people say what would have been unthinkable in the years when the Soviet occupation was a synonym for brutality: that the 'Russian time,' as it is known, was not so bad after all, at least in Kabul."[46]
>
> "Aysha, who like many Afghans uses only one name, sobbed as she clutched the young man's hand. 'The people who did this are no Muslims,' she said. 'They are the henchmen of Satan and they will surely suffer in hell."[47]
>
> "'I cannot see any hope,' she said today [after the Taliban forces overran Kabul]. 'When I look around me now I see nothing but a new calamity, a life of further misery on top of everything else we have suffered.'"[48]

Eventually, a popular base of support becomes available to political leaders with the courage and ability to pursue a peace settlement. The longer the war or conflict goes on without resolution, the greater become hostile attitudes towards the war. Breakthroughs, whether in Israel/Palestine, Northern Ireland, Guatemala, Mozambique, Angola, or El Salvador, have their foundation in the popular hatred of war. In contrast to the war, the work of the ICRC demonstrates that life can be safer, healthier, more productive, and full of compassion.

6

Guatemala

The intensity of strife in Guatemala has risen and fallen during the past forty-two years. The situation is one of classic, class-ethnic violence. The more affluent *ladinos* have killed to keep control of the government. Maya Indians, *campesinos,* and urban workers led by intellectuals and labor leaders have struggled to have a government that would serve their needs. The oppressed spawned a guerrilla force, and it killed too. Perhaps 150,000 died—mostly Indians. Ten times as many were displaced, of whom about 40,000 fled across the border into Mexico.

I arrived in Guatemala City on March 19, 1996, trying to get a closer look at the war, see old friends, and interview ICRC delegates. Guatemala City is the home of ICRC's regional delegation for Central America and the Caribbean, and is managed by delegates Patrick Zahnd, Graziella De Vecchi, and Laurent Burkhalter.

One day after my arrival, the guerrilla movement, the Guatemalan National Revolutionary Union (URNG), declared a cease-fire. Newly elected President Alvaro Arzú, who had met rebel leaders in Mexico City in February—the first president to do so—then ordered the army to halt all operations the next day. That the suspension of hostilities occurred within days of my arrival was purely coincidental, although, having studied the war for decades, I was struck by its considerable irony.

The people of this beautiful country began to contemplate an end to its long period of violence and suffering. It became possible to imagine that the bravery of progressive leaders on all sides would finally overcome the brutality of more reactionary leaders on all sides. The courage of the peacemakers won them a fragile ascendancy. Third-party political and humanitarian intervention, most notably by a Group of Friends (mediators from Norway, the United States, Mexico, Venezuela, Spain, and Colombia), the United Nations Human Rights Observer Mission (MINUGUA), the staff of the United Nations High Commissioner for Refugees (UNHCR), UN mediator Jean Arnault, the ICRC, the Archbishop's Human Rights Office, the Lutheran World Federation, the observers of Human Rights Watch/Americas, and a growing number of like-minded organizations, was key in winding down the war.

On December 29, 1996, before a capacity crowd filling the central square in Guatemala City, the government and the URNG signed the Accord for a Firm and Lasting Peace. This day of both solemnity and happiness marked what most Guatemalans hoped would be the end of decades of violence.[1] Previously, agreements had been signed on human rights, Indian rights, poverty and land reform, constitutional and electoral reform, the renunciation of violence, and the reintegration of the guerrillas into Guatemalan society.[2]

The ICRC's role in helping to resolve the war and, furthermore, its efforts to make the war distasteful in Guatemala's historical memory, have been specifically tailored to unique conditions in Guatemala. The ICRC's unspoken mission, as with its activities in other internal wars, is pursued behind the operation of the main mission of Geneva Convention–mandated relief and protection given to the ICRC. As will be explained, the ICRC's strategy to end the war and keep it from rekindling relies preponderantly on the dissemination of international humanitarian law. The strategy's effectiveness, non-political facade, patient blending with the actions of other third parties, and strong humanitarian foundation further add to the ICRC's reputation for peaceful service in the service of peace.

Not all third parties were always on the side of peace.

The United States government, in particular, had blood on its hands. Its intervention in 1954 largely created the conditions for many a premature death. Since then, U.S. foreign policy toward Guatemala became two policies: One—the formal, official policy of the White House, State Department, and Congress—moved toward political reform, human rights, and ending of the war. The other—that of the bureaucratic processes of the Pentagon and CIA—created a subterranean policy that served to perpetuate the war.

THE ORIGINS OF THE WAR

Some would argue that the conditions for an internal war were created early in Guatemalan history. Spanish colonization policies deliberately bifurcated society. Control of land would facilitate control of government and people, and land went to the Spanish and their *ladino* offspring. Large plantations, or *fincas* consumed the arable land. Their crops, first coffee and cotton and then bananas, would be for export. The Indians, descendants of the Maya, would labor in the fields and build the cities virtually as slaves. When harvests were in, the Indians would go back to their remote villages and continue a life filled with colorful costumes, amalgamated Catholic-Mayan religion, large families, poverty, distinctive local languages, illiteracy in Spanish, and disease. Separation from Spain in 1821 hardly changed the social and political structures in Central America.

Coffee boomed early in this century, and it was joined by a new crop and a new player in Guatemalan society. The United Fruit Company bought what eventually amounted to over half a million acres of the country's farmland. United Fruit quickly earned the name *El Pulpo* (the octopus) as it created a banana empire. It built the major harbor and constructed or took over the country's railroad system. Another American company owned and operated the major electric utility company. United Fruit worked in tandem with U.S. diplomats to maintain a cozy relationship with the Guatemalan oligarchy. This team, together with U.S. coffee interests, forced the expulsion of German-Guatemalans during World War II. At one time, German-owned plantations had produced half of the coffee production in Guatemala.

The development of a modern export sector and the infrastructure to support it gradually created a political class that was hostile to the oligarchy. Surging economic production during World War II swelled the ranks of the (albeit still small) professional and business middle class, urban and rural workers, and young army officers, as well as the educated leaders each group produced. Wartime production did not result in widespread prosperity. In spite of increased economic activity, prices for Guatemala's agricultural commodities stabilized or fell. The war cut off its European markets, making its exports even more dependent upon the U.S. market. And that market set prices. As a result, the war created a diverse modern sector that became increasingly politicized when it failed to receive its economic due. This new middle and working class was also barred from political participation. In 1944, without U.S. opposition, a progressive *ladino* coalition overthrew the reactionary regimes of Jorge Ubico and his short-term successor, Juan Frederico Ponce, held free elections and installed the liberal government of Juan José Arévalo.

President Arévalo moved left, consolidating his political base of urban workers and intellectuals, agricultural workers, small businesspersons, and *campesinos.* He referred to his ideology as "spiritual socialism." Unions and their right to collective bargaining and to strike were legalized under the 1947 Labor Code. *Campesinos* could rent unused land from the large landowners. A social security system was established, education was expanded, and health services were brought to rural areas.

This democratic revolution was fertile ground for a growing communist movement. A liberal government made the realization of communist ideals more likely, as progressive legislation sought to narrow the wide rich-poor gap. Progressive political leaders raised the expectations of workers and peasants. If their government was for them, how could the old elite continue to exercise power over them? Notions of equality flourished amid conditions of gross inequality. The elite, with 2.2 percent of the population, owned over 70 percent of the land, and a quarter of relatively well-off *ladinos* owned another 20 percent, which left about 10 percent of the land to the rest of the population.[3] In 1950, Communists won control of the largest labor union.

And so began the tragedy. American Cold War–driven national interests collided with the reform and modernization in Guatemala. The collision would shape Guatemalan politics and feed a thirty-five-year guerrilla war.

The FBI began warning of the growing Communist influence in Guatemala as early as 1946. These warnings and other Cold War alarms convinced President Truman that Arévalo was a Communist. Still, U.S. officials did not interfere in the election of Jacabo Arbenz Guzman in 1950, expecting him to be less radical than Arévalo. They were mistaken.

President Arbenz boldly initiated a series of domestic reforms that widened Washington's perception that Guatemala was moving towards the Soviet side in the Cold War. Arbenz also began a pattern of anti-American foreign-policy moves, some quite brainless, that solidified that perception.

Domestic reforms inevitably had to injure the interests of U.S. corporations operating in Guatemala, principally the United Fruit Company. Arbenz needed more revenue and more land for his reforms. New social services required revenue. Increasing the holdings of *campesinos* in order to serve their needs and to grow food for the domestic market required a rapid expansion of state-controlled land. These resources were largely held by Americans.

Arbenz first tried to end-run American enterprises to produce revenues. He began a hydro-electric project to compete with the U.S. electric company, a highway system to compete with the railroads, and a new port to compete with United Fruit's Puerto Barrios. In 1951, he went further, nationalizing the railroads and electric company. Arbenz brought suit against foreign corporations to collect unpaid taxes.[4] To gain land, Arbenz made available state-owned land for distribution and targeted the vast holdings of El Pulpo. United Fruit held fallow most of its land in order to maintain its fertility, avoid disease, and not flood the market with bananas. Arbenz called such land "idle" and provided for its expropriation under his unanimously passed Agrarian Reform Law of July 1952. About 400,000 acres of United Fruit's 550,000 acres were eventually targeted for a payment of $1,185,000 in 25-year, 3-percent-interest government bonds.[5] The proposed payment would come to less than $3.00 per acre. The dollar amount equalled United Fruit's estimation of the land's value in its tax record. It was unfair, United Fruit argued with some justification, for it to give up land priced at an amount that the law allowed the company to set for tax purposes only. Regardless of the popularity of bashing United Fruit, Arbenz's move was as unwise as it was unfair. It brought about a strong letter of protest from the U.S. State Department and helped solidify the Truman administration's conclusion that the problem in Guatemala was communism. Truman's successor, Dwight Eisenhower, came to the same conclusion.

A series of foreign-policy moves by the Arbenz government then led directly to Eisenhower's decision to intervene. Arbenz continued Arévalo's somewhat romantic support of a largely paper, anti-caudillo Caribbean Legion, which mainly sought to overthrow Nicaragua's dictator Anastasio Somoza. More seriously, Arbenz succeeded in getting Guatemala's Communist Party legalized in 1952, and the subsequent election saw four of its members elected to the legislature in Guatemala City. When Stalin died in March 1953, this congress observed one minute of silence in his memory. Arbenz appointed Communists to his administration.[6] The "clincher," as American officials would say in supporting the decision to overthrow the Arbenz government, was Guatemala's covert operation in ordering Soviet bloc arms. The arms sailed into port in May 1954. Previously, Truman had stopped arms shipments to Guatemala, and Arbenz believed he would need more arms than he had to protect his government.

While true, the very fact that he ordered them created the need to have them. The Soviet bloc arms triggered U.S. armed intervention.

Although domestic reforms at the expense of United Fruit brought protests from Washington, it is doubtful that reforms alone would have brought U.S. intervention. Under President Lazaro Cardenas in the late 1930s, Mexico had been far more radical in expropriations than Guatemala, but without creating the threat of intervention. The key for Mexico, as World War II approached, was fidelity to U.S. foreign policy in its growing opposition to Nazi Germany; in the Cold War it would be fidelity to American leadership of the "free world" against international communism. The contrasting Bolivian experience confirms that communism, and not domestic reform at the expense of U.S. corporations, was the trigger for U.S. intervention in Guatemala. At the same time as the Guatemalan domestic reforms, Bolivia was engaged in similar policies. But the Bolivian government was staunchly hostile to its country's Communists, so instead of intervention, it received generous U.S. financial assistance.[7]

In 1954, the American press, Congress, and the president remembered well how just a few years earlier the Chinese Communists had been called "agrarian reformers," and how Mao then took China into the Soviet bloc after the Communist victory in 1949. The "free world" shrank. They did not want this to happen in "our backyard," the domain of the Monroe Doctrine. The Monroe Doctrine was really a declaration that established a sphere of influence. And spheres of influence designate the area from which vital security interests can be threatened. For the United States, that area was the Caribbean and Central America, the only land/sea avenues by which an enemy could most directly access the nation and interfere with its commerce. If Guatemala were to be lost to communism, then a base of subversion would be created to spread the "red tide" throughout the area. Eisenhower and Dulles were acutely aware that the poverty of the masses and a general hatred of the "colossus of the North" constituted a fertile breeding ground for left-wing revolutions, and that the Soviet Union could supply the ideological and organizational model for such revolutions and then arm them and back the revolutionaries as allies.

Whether Arbenz was a Communist or not was not the issue for Eisenhower. "His actions," wrote Eisenhower in his memoirs, "created the strong suspicion that he was merely a puppet manipulated by Communists."[8] Eisenhower's newly appointed ambassador, John E. Peurifoy, went further. After a six-hour meeting with Arbenz, Peurifoy reported to Secretary of State Dulles in late 1953 "that unless the Communist influences in Guatemala were counteracted, Guatemala would within six months fall completely under Communist control."[9] Communist control, under the assumptions of the time, meant that, since Moscow firmly controlled international communism, Moscow would control Guatemala. "Something had to be done quickly," Eisenhower reflected, and the first task involved a propaganda barrage against Arbenz.[10] The stage for doing so was the Tenth Inter-American Conference of the Organization of American States, which met in Caracas in March 1954, and at which Dulles introduced a draft resolution. Entitled a "Declaration of Solidarity for the Preservation of the Political Integrity of the American States against International Communist Intervention," it was approved by a vote of 17 to 1 (Guatemala), with Mexico and Argentina abstaining and Costa Rica absent.[11] The Soviet bloc's arms shipments in May, in conjunction with Arbenz's declaration of a state of siege and the arrest of opponents to his government, finalized Washington's decision to launch Project PBSUCCESS.

Much has been written about the covert operation to overthrow Arbenz, including Richard H. Immerman's notable study. Most writing has been highly critical of the United States because it instigated a very nasty, very long internal war. While this is true, some of the blame must lie with Arbenz. His failure to realize that the threat of defection (or the opportunity to promote a defection) produced superpower intervention in *every* Cold War confrontation—whether in France, Italy, East Germany, Czechoslovakia, Korea, Indochina, or Iran—is inexplicable. When *The New York Times* in 1954 editorialized that Guatemala was "this spot of Communist infection," Arbenz should have realized that his choices were to either purge the Communists or face overthrow.[12] Of course it wasn't fair or legal or just. The Cold War transformed the superpowers into self-righteous bullies as they intervened around the globe to support their clients and to subvert those of the other side. Sovereignty, the principal upon which the nation-state system rests, suffered mightily.

CIA-led Project PBSUCCESS was indeed a stunning Cold War success. The CIA recruited an exiled Guatemalan colonel, Carlos Enrique Castillo Armas, and Armas recruited the CIA as his ticket to power. With a handful of men, Armas crossed the Honduran border and encamped while the CIA broadcast over its radio *Voice of Liberation* bogus reports of stunning victories for the rebel forces. CIA aircraft strafed Guatemala City. The operation was mainly psychological. Its object was to panic Arbenz so as to drive a wedge between him and the army, and panic he did. Revealing his distrust of the army, Arbenz asked the army to arm workers and peasants. The army refused. Having lost the authority of his office, Arbenz resigned and fled. In triumph, Ambassador Peurifoy put together a junta and rigged an election that brought Armas to power in October 1954.

The free world was saved. After the overthrow, Dulles commented: "Now the future of Guatemala lies at the disposal of the Guatemalan people themselves."[13] Somehow, once the threat of communism was eliminated and some American foreign aid provided, everything would be all right. But it was not. The intervention had long-lasting effects.

THE CONTEXT OF THE WAR

The principal legacy of Arbenz's overthrow was the militarization and nationalization of Guatemalan politics. The military became the power base of the elite-run *ladino* government, and, as right-wing militaries are wont to do, they used the gun to keep control over the lower classes. This produced among the victims of government repression a Marxist worker-peasant guerrilla response. The job of putting down the "communist" guerrillas then gave the army even more political power, as well as more military aid and support from the United States.

Eventually, the war encompassed the whole country, and this spread politics to every village and neighborhood. For the first time, *ladino* and Mayan were integrated on a full-time basis, even though they killed each other regularly. *La violencia* increasingly attracted third-party intervention in Guatemalan politics, with the United States moving away from *carte blanche* support of the government and toward policies that favored reform and the war's resolution. This eventually put official U.S. policy in line with UN and NGO efforts. However, at the same time, the Pentagon and CIA kept their connections with and continued their support for Guatemalan military and police personnel. These bureaucratic connections hampered third-party efforts to stop the murders, disappearances, ambushes, forced migrations, and executions. But, as it did elsewhere, the end of the Cold War calmed left-right (but not religious or ethnic) struggles and helped promote the March 20, 1996 cease-fire and the December 29, 1996 final settlement.

Proof that the military is the center of national politics does not come from having military officers in the presidential palace. America, after all, had Washington, Jackson, Grant, McKinley, and Eisenhower in the White House. Instead, the proof comes when the military establishment can select and dispose of chief executives at will. That was the case in Guatemala until 1995, and many still doubt that everyone in the military has given up that kingmaking role. Having the chief executive as a captive left the military free to make its own policies. It did this in coalition with the economic elite.

Soon after the overthrow of Arbenz, the Labor Code was repealed; land was returned to United Fruit, "subversive" political parties were banned; and thousands of communists, liberals, students, teachers, journalists, union leaders, and Mayan activists were executed without trial.

Elections were still held, although the franchise was severely restricted. However, whenever segments of the military were unhappy with the leadership, whether civilian or military, they either organized a coup or made a deal to maintain the president in return for his pledge to obey their most powerful faction. Liberal junior officers attempted coups in 1960 and 1962. They failed, and some of their leaders took to the eastern highlands to begin guerrilla operations. Coups were successful in 1963 (supported by Washington when President Ydígoras had the courage to allow Juan Arévalo to return to Guatemala to campaign for president), 1982, 1983 (also backed by Washington, but this time to remove an anti-democratic military caudillo), and 1993. Civilian presidents, such as Julio César Méndez Montenegro (1966-70) and Vinicio Cerezo (1985-90) had their progressive reforms thwarted when they were brought to heel by the military.

During this time, the anti-guerrilla campaign ebbed and flowed, backed by U.S. military equipment and advisors. In the 1960s, for instance, nearly a thousand U.S. Green Beret Special Forces operated alongside the Guatemalan army. Twenty-eight Green Berets died.[14] Private, anti-communist armies organized by the rural oligarchs were also allowed to operate. In Guatemala City, death squads killed suspected guerrilla sympathizers with impunity. In the countryside, the army faced guerrillas that numbered only several hundred, and reached a maximum of perhaps six thousand in the 1970s. For example, the Castroite Yon Sosa, leader of MR-13 (the Revolutionary Movement of November 13, the 1960 date of the failed coup), had only a hundred or so guerrillas under his command when he was killed in 1970.

The small size and mobility of the rebels and the rough terrain in the highlands made eradicating the guerrillas extremely difficult. Coordination among the guerrilla groups—they combined into the Guatemalan National Revolutionary Union (URNG) in 1981—also made their destruction difficult.

A Town Square in Guatemala.

Four heavily armed police return to barracks after a patrol of the town. Anytime you see police that look like soldiers, chances are that they are soldiers and the country is in some state of war with itself. The ICRC sees a world with too many guns, especially small arms, that get in the hands of too many people. It calls for restrictions in the trade of these weapons. *(Photo by Nicholas O. Berry)*

The army and the private militias, therefore, adopted the strategy of terrorizing the guerrilla's base of support: the peasants and Indians. During the presidency of General Fernando Lucas (1978-82), for example, one estimate numbered over 100,000 Guatemalans killed or "disappeared," and more than 1 million displaced.[15] Entire villages and all the villagers disappeared.

In a 1996 report, the NGO Human Rights Watch/Americas provided a retrospective on the army's strategy, subtitled: "Background: Setting the Stage for Violence."

> Tens of thousands of Guatemalans fled to southern Mexico between 1980 and 1983, as the army systematically massacred and razed entire villages, primarily in the departments of El Quiché, Heuhuetenango, the Petén, Alta and Baja Verapaz, and San Marcos. Hundreds of thousands more were displaced internally. To consolidate its control in the countryside, the army relocated tens of thousands of displaced Guatemalans into centralized "model villages," designated strategic areas as "development poles," and required villagers to form civil patrols. Much of the Ixcán municipality, including the Zona Reyna in northern Quiché province, formed part of a strictly regimented development pole. The establishment of model villages and development poles enabled the army to control and coordinate assistance to rural populations as part of its counterinsurgency strategy. The army organized, trained, and armed civil patrols in hundreds of villages to augment its local presence and control. Although the 1985 Constitution made patrol duty voluntary by law, participation in the patrols often remained obligatory in practice. Civil patrol involvement in human rights violations remains among the principal human rights problems in Guatemala.[16]

In essence, the army's strategy focused on noncombatants (civilians) in an effort to break their links with the elusive guerrillas. Killings, forced relocations, puppet civil patrols, and government-sponsored social-assistance and economic-development projects (which demonstrated that the government could and should help the Maya) consciously targeted civilians. As in other internal wars, over 90 percent of the dead were civilians.

These excesses brought periodic protests from Washington. President Carter ordered an arms embargo in 1977. During his administration, one State Department official lamented: "What we'd give to have an Arbenz now. We are going to have to invent one, but all the candidates are dead."[17] President Reagan resumed military aid, but President Bush renewed the embargo after a series of outrages by the Guatemalan army in 1990. Ten academics from the University of San Carlos were murdered; about a hundred Indians were

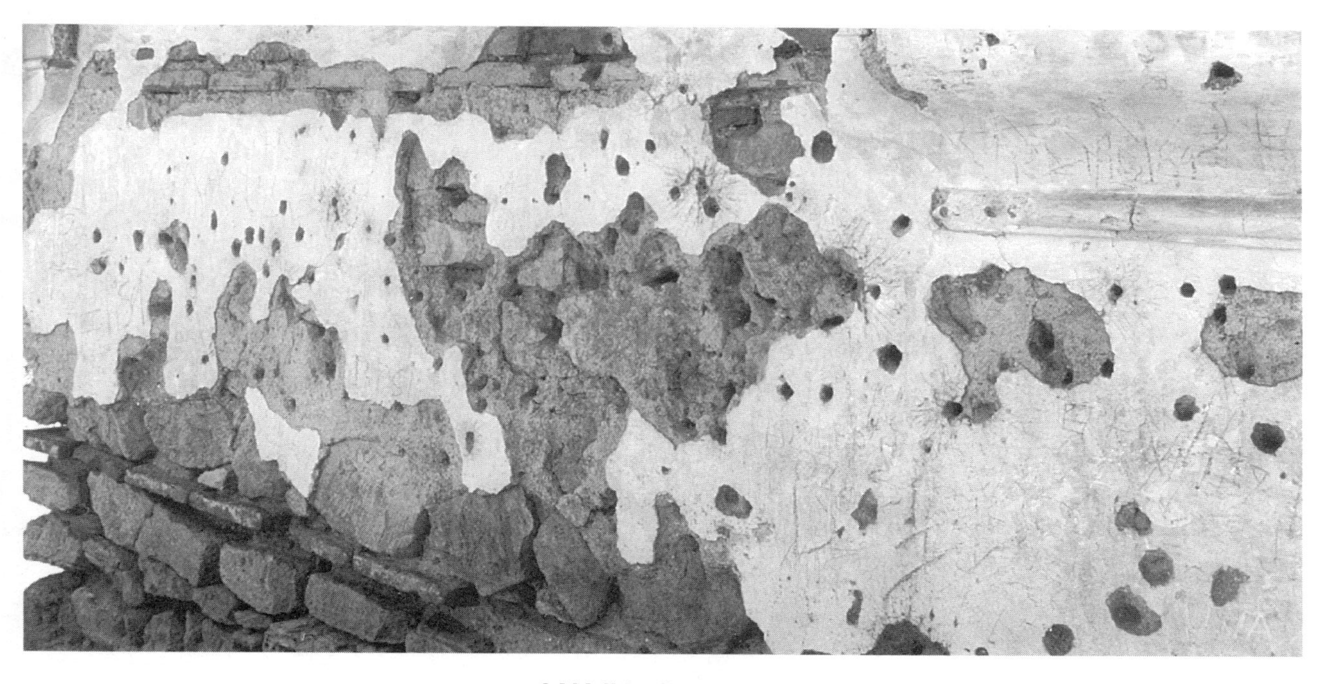

A Wall in Guatemala.

One of the many walls used either for firing practice or in actual executions. I have seen similar walls in Italy, Greece, Libya, and Poland—eerie reminders of the all-pervasive violence of war. *(Photo by Nicholas O. Berry)* massacred at Santiago Atitlán; an American owner of a rural inn, Michael Devine, was hacked to death on suspicion of aiding the guerrillas; and Dianna Ortiz, an American nun, was kidnapped and raped by CIA-supported troops before being rescued by an American CIA operative.

In April 1996, I commented to a CIA official who had major operational responsibilities in Central America: "You really recruited a bunch of unsavory characters in Guatemala, didn't you?" He replied with candor: "Who else was there?"

THE ROAD TO RESOLUTION AND THE ICRC

The winding-down of the war was and remains a team effort. The forces of reconciliation in Guatemala were supported by an array of foreign-based third parties. Each had particular roles; each could read the war, see what the other third parties were doing to stop it, and then coordinate and harmonize their efforts to end the violence without overtly coming together (which would create the impression of massive foreign intervention). The ICRC, in particular, played a crucial role.

Conflict resolution, to reiterate, requires that warring parties are able to negotiate—that they are either under strong leadership or are so weak that their leaders have no choice but to negotiate. When such parties agree to meet, they often require third-party mediation to reach a settlement that gives everyone more benefits and fewer costs than the benefits/costs of continuing the conflict. In other words, the warring parties must be able and willing to negotiate a settlement, a settlement that has a high probability of being their most beneficial option.

Except for the Guatemalan army and segments of the economic elite, everyone else—journalists, scholars, NGOs, the ICRC, IGOs, and foreign governments—has known who was the singular major impediment to ending the war. Everyone knew and still knows that this party must be induced to drastically reduce and change its political role. The Guatemalan army was and is the key to peace.

In 1989, soon after the Cold War ended, a strategy was crystallized. An uncoordinated but concerted effort by both domestic and foreign antiwar participants sought to restrain and professionalize the Guatemalan military and their civilian allies. Weakening any target in politics usually means dividing it and then supporting the good side and opposing the bad. The good side in the Guatemalan army are those officers who want to stay out of politics. The first task was to divide the military into good and bad. Then it would be possible to support the good side, the side that would treat civilians according to law and allow civilians to be the government's political base. A political base of democratic government is made up of people who can choose and remove government officials through regular elections among contenders for office. The dominant, good faction of the army had to be induced to give up its kingmaking role. This would leave the civilian sector in political ascendancy. What is called a "civil society" would exist.

The essential and ultimate requirement was that the army give up control of its Mobile Military Police (PMA). Since control of a police force is essential for law enforcement, a controlling party can determine what laws get enforced. In other words, military leaders are the law-makers. In democracies, the party that controls the police is the civilian team of elected legislative and executive leaders. Combined, they are the law makers. Unlike the army, these lawmakers are responsive to and dependent upon civilian voters.

A politically weakened, reformed, and more professional army would be more willing to let civilian negotiators negotiate.

In summary, three tasks faced the peacemakers.

1. Professionalize the military, including: a) penalizing the army and its elite supporters for transgressions; b) monitoring its operations; c) reducing CIA and U.S. military backing; and d) inducing it to treat civilians as if they had human rights.
2. Build up civil society, including: a) strengthening civilian organizations in both the public and private sector; b) protecting Indians and reform leaders; and c) inducing the URNG to give up terror and guerrilla operations and thus committing it to a negotiated settlement in harmony with reform leaders in civil society.
3. Provide third-party mediation, including: a) processing individual human rights complaints; and b) bringing the representatives of the warring parties together to negotiate.

Before we analyze those who worked on all these tasks, especially the ICRC, the reader should keep in mind the importance of the end of the Cold War in facilitating the chance for peace. When the Cold War disappeared, the international aspect of the URNG rebellion also disappeared. The URNG could no longer be attacked as the handmaiden of the Soviet Union. The Guatemalan army thereby lost most of the ideological foundation of its repression. No longer could it claim to be fighting international communism. This transformed the war into a domestic conflict between Guatemalans, and surely Guatemalans could come to some kind of settlement. The conflict became more of a family affair.

The end of the Cold War also helped transform the relationship between the Guatemalan army and the U.S. government, especially the Pentagon and

CIA. Washington no longer needed the Guatemalan army to be on the front line against international communism. In fact, the army's repressive behavior increasingly became an embarrassment, to the point where the dual foreign policies of the U.S. government eventually merged on the side of ending the war by negotiation.

Many organizations worked on professionalizing the Guatemalan military. Penalizing its transgressions fell mainly to the U.S. government.

As already described, Washington cut off overt security assistance to Guatemala after the outrages of 1990. The CIA and Pentagon specialized aid continued, however, mainly through anti-drug programs (even though members of the Guatemalan army allegedly had a piece of the drug trade). A turning point in diminishing the political power of the Guatemalan army came in 1993. Amid a growing clamor for reform and of increasing newspaper reports of government corruption, an unpopular President Jorge Serrano Elias tried to repeat what Peru's president Alberto Fujimori had successfully pulled off two years before. On May 25, 1993, Serrano attempted an *autogolpe* or self-coup. He suspended some provisions of the constitution and disbanded the congress, Constitutional Court, and Supreme Court of Justice.[18] He acted in collusion, of course, with the most reactionary factions in the military. The Clinton administration acted quickly. It threatened to impose sanctions on the economic elite—restrictions on trade, economic aid, and international financial assistance—unless it could convince the army to withdraw support for Serrano's *autogolpe.*[19] Several European governments supported Washington's threats, and the administration was bolstered by Guatemalan journalist, labor, academic, and business organizations favoring reform. The army backed down, Serrano resigned, and this set in motion a series of events that demonstrated the growth of the civilian sector's power. One event in particular showed this trend. On October 5, 1995, a detachment of soldiers massacred 11 villagers and wounded more than 30 others in the village of Aurora 8 de Octubre, in Xamán, after villagers protested the soldiers' presence on community lands. President Ramiro de León Carpio, who was appointed president after Serrano's forced resignation, felt sufficiently strong to take action. He appointed a special commission to look into the case, accompanied it to Xamán on October 6, and accepted his government's responsibility for killings. President de Leon pledged to bring those who committed the crime to justice. He dismissed the area military commander and three days later accepted the resignation of the defense minister, General Mario René Enriquez.[20]

President de León's special commission was backed up by an array of domestic and international monitors. Within hours, representatives of the Archbishop's Human Rights Office (ODHA) and the Rigoberta Menchú

Foundation were in Xamán. So too were officials from the UN High Commission for Refugees (UNHCR) and the year-old United Nations Verification Mission (MINUGUA). (MINUGUA was formed after the government and URNG signed a human rights accord in March 1994, as discussed later in this chapter.) Other human rights personnel and journalists soon followed.[21] It is difficult to fight a war against civilians with human-rights monitors on the battlefield or available on a moment's notice. Commanders and their soldiers, if they know they will be prosecuted for war crimes (and this is still problematic in Guatemala), will hesitate to commit war crimes.

To repeat, stopping ugly internal wars is a team effort. Some unusual people joined the team in Guatemala. Three American women did so, and they are largely responsible for diminishing the CIA's backing of Guatemala's security forces and increasing White House and congressional pressure on Guatemala to reform.

Sister Dianna Ortiz taught literacy to Mayan children, for which she was kidnapped, tortured, and raped in November 1989. Rescued by a CIA agent whom the kidnappers called "their boss," Sister Ortiz returned to the United States to learn the truth about her abduction.[22] Her fruitless inquiry culminated in a hunger strike and vigil outside the White House in April 1996. This produced an interview with Hillary Rodham Clinton, who promised to help.

Sister Ortiz was joined in her vigil by Jennifer Harbury, an American lawyer whose guerrilla husband, Efrain Bámaca, was captured by the military in 1992 and eventually executed without trial.[23] Back in Guatemala, Carol DeVine pressed for information about the army's murder of her husband, American innkeeper Michael DeVine, in 1990.[24] The Bámaca and DeVine cases were taken up by Representative (now Senator) Robert G. Torricelli (D, NJ) in 1995 when he accused a CIA-paid Guatemalan officer, Colonel Julio Alpírez, of responsibility for both deaths.[25] Soon after the accusations, President Clinton ordered the Intelligence Oversight Board (IOB) to investigate the deaths and disappearances of Americans in Guatemala and any involvement by U.S. agencies. Even before the release of the IOB's report, two high-level CIA officials were fired and nine others disciplined.[26] The classified report, delivered to President Clinton on June 28, 1996, found "unacceptable" the CIA's use of informants who were involved in assassinations, torture, kidnapping, and murder. None of the informants referred to remain on CIA's payroll, and CIA Director John M. Deutch had already issued a directive barring the recruitment of unsavory informants unless clearly in the national interest.[27] Congress again cut off all aid to army and security forces in the fall of 1995.

The grizzly information on all the cases filtered back to the Guatemalan media and became big news. Jennifer Harbury used the internet extensively to let her supporters and the media know of the progress in the various cases.[28] The result was a further weakening of the military's and security forces' reputations to the point where President Arzú felt secure enough in February 1996 to cashier Colonel Alpírez and others in the military.[29] Even the military command realized that some very bad apples were spoiling the entire barrel.

Besides weakening the military in relation to civilian authorities, the concerted effort to end the war in Guatemala included an essential task. This task was the primary responsibility of the ICRC—get the military and police to abide by international humanitarian law. Monitoring their activities and weeding out their principal human-rights violators was a good beginning. But for long-lasting reform, it was essential to train Guatemala's armed forces to understand not only the rules of international humanitarian law but also how following the rules could help in their operations and reputation. This mission was fully legitimate under the ICRC's Geneva Convention rights to assist in wars. I will make the case that, in effect, the ICRC was (and is) doing the army and police a big favor.

What made the ICRC concentrate on the dissemination of international humanitarian law was the relationship between the warring parties after the return to civilian government in 1986. The UNRG, militarily weak, saw potential allies in the non-revolutionary civilian sector, most notably in the Catholic Church and the business community. The church sponsored a National Dialogue for peace in 1989, and a large business organization entered the Dialogue the next year.[30] In 1991, the Serrano government opened talks, although the army insisted that URNG unilaterally ceasefire. It then added that the guerrilla forces must disarm before it would seriously negotiate. The failed *autogolpe* of 1993 and the outrages by the army brought new impetus to the peace talks. In March 1994, a Global Human Rights Accord was signed with the backing of all segments of society except the army and sections of the agricultural elite. It brought in UN officials for monitoring human rights and for mediation.

The ICRC and everyone else knew, therefore, that the key to a peace settlement depended upon the military and police. The army's continued repression of Indians and civilian reformers was the only reason the war continued. The URNG wanted to settle. What made restraining the armed forces even more pressing was a 1994 accord that provided for the resettlement of displaced persons, including those who had fled to Mexico. The discovery of oil deposits on resettlement lands added a new motive to the war between the elites and the Indians. As ICRC head delegate, Patrick Zahnd, told me: "To invite the refugees back when the war is not over makes it very difficult [to protect civilians]."

The more that the military understands and respects international law, the ICRC believes, the more the political process will dominate the conflict. This made the dissemination of international humanitarian law "vital," according to ICRC delegates: "We are not a relief agency; we are a protection agency." And the way to protect civilians was not, as I had expected before arriving in Guatemala, to be in close contact with and providing relief services to the Indians, but to be in the closest contact with the government, military, and police forces. The ICRC offers only the bare minimum of field operations, which mainly involves visits with detainees, an orthopedic program for mine victims, and small health projects.

The first task of the three Guatemala City ICRC delegates was to gain legitimate access to government and military officials. For this, the ICRC negotiated and signed a protocol with the government in 1987 that gave the delegates the equivalent of diplomatic status. They received strong backing from Geneva. In March 1995, for example, the president of Guatemala visited ICRC headquarters and was informed of the ICRC's readiness to provide for the protection of civilians and detainees in Guatemala. Once access was assured, it was necessary for the delegates in Guatemala City to get an agreement that international humanitarian law be taught in all military schools and police academies and that these schools all have legal advisors. This took time and patience, and by 1995 law was taught even at military installations in the field. One delegate now spends 80 percent of his time on disseminating international law in these schools, while the others make arrangements and develop materials.

The curriculum has three objectives.

The first is to convince more and more military and police officials that international humanitarian law applies to the internal war in Guatemala. If accepted, then the laws of war thereby become legitimate for guiding military behavior. Common Article 3 of the four 1949 Geneva Conventions and the Additional Protocol II of 1977 specifically cite this norm. This prescription of Article 3 and Protocol II is the first subject covered. Article 3 says: "In the case of armed conflict not of an international character occurring in the territory of one of the High Contracting Parties, each Party to the conflict shall be bound to apply, as a minimum, the following provisions." The 28 articles of Protocol II of 1977 then elaborate the protection given to civilians in civil wars as provided for in Article 3. Guatemala ratified the Geneva Conventions, with minor reservations, on May 14, 1952, and did the same with Protocol II on October 19, 1987.[31] In addition, the ICRC proclaims in all its publications concerning internal wars that Article 3 must now be regarded as part of customary international law.[32]

Once political and military officials accept the fact that the Geneva Conventions and Additional Protocols apply to "non-international armed conflict of high intensity," the second objective of the curriculum is to inculcate the legal provisions that apply to these wars. Article 3 continues:

1. Persons taking no active part in the hostilities, including members of armed forces who have laid down their arms and those placed hors de combat by sickness, wounds, detention, or any other cause, shall in all circumstances be treated humanely, without any adverse distinction founded on race, colour, religion or faith, sex, birth or wealth, or any other similar criteria.

   To this end, the following acts are and shall remain prohibited at any time and in any place whatsoever with respect to the above-mentioned persons:

   a) violence to life and person, in particular murder of all kinds, mutilation, cruel treatment and torture;
   b) taking of hostages;
   c) outrages upon personal dignity, in particular humiliating and degrading treatment;
   d) the passing of sentences and the carrying out of executions without previous judgment pronounced by a regularly constituted court, affording all the judicial guarantees which are recognized as indispensable by civilized peoples.

2. The wounded and sick shall be collected and cared for. An impartial humanitarian body, such as the International Committee of the Red Cross, may offer its services to the Parties to the conflict.

   The Parties to the conflict should further endeavour to bring into force, by means of special agreements, all or part of the other provisions of the present Convention.

   The application of the preceding provisions shall not affect the legal status of the Parties to the conflict.[33]

Protocol II extends these provisions. For example, Article 6 details the requirements for a properly conducted trial, and Article 5 prescribes a code for the treatment of people in custody. Protocol II also protects cultural objects, places of worship, and "works and installations containing dangerous forces" (such as dams, dikes, and nuclear power stations), and prohibits the forced movement of civilians.[34]

This body of international humanitarian law is designed to protect the civilian population who are, in overwhelming numbers, the victims of internal war. On this basis, ICRC delegates in Guatemala rightly say that "We're not a relief agency; we're a protection agency."

We also see the "right of initiative" included in the common Article 3 whereby the ICRC is specifically named as a legitimate provider of services to the parties in armed conflict.

In Guatemala, one service—disseminating the laws of war—depends upon reaching as many people as possible inside the government and armed forces. ICRC delegates talk to the writers of law in the executive and legislative

CATALOGING IN PUBLICATION 12/03

ACSM's health-related physical fitness assessment guidelines / editors, Gregory B. Dwyer and Shala E. Davis. Baltimore, Md. : Lippincott Williams & Wilkins, c2004.

p. ; cm.

Companion v. to: ACSM's guidelines for exercise testing and prescription / American College of Sports Medicine ... [et al.]. 6th ed. c2000.

Includes bibliographical references and index.

ISBN 0-7817-3471-1

(Continued on next card)

2003060617

MARC CIP 12/03

[0309]

Teaching International Humanitarian Law (IHL) to Soldiers.
The ICRC assists military forces in disseminating IHL, producing for enlisted soldiers picture books showing how to behave toward prisoners, the wounded, and civilians; and how to respect medical and relief workers operating under the red cross and red crescent symbols. If civil war is "civilized" to the point that killing becomes difficult, then war will find itself a less used institution. *(Photo courtesy of ICRC, by Paul Grabhorn)*

branches so that international humanitarian law gets incorporated into domestic law. These discussions also take place in all state governments. In the military, instructors are trained in a forty-hour course to supplement and eventually take over the teaching of the laws of war. Officers are sent to the International Institute of Humanitarian Law at San Remo, Switzerland for advanced instruction. More than a dozen had attended by 1995.

The ICRC held a seminar for all of Latin America in September of 1995 in Guatemala City. Two representatives from nineteen countries studied the laws of war, with results "that helped a lot" to legitimize and clarify the rules that restrain combat. The seminar also addressed UN peacekeeping operations, the prevention of war crimes, and the ICRC's campaign against land mines.

Senior Guatemalan officers are given manuals on the laws of wars during their sessions with ICRC delegates. The manual uses case studies, such as those on the treatment of prisoners. One delegate interviewed had just returned from a four-month, 12,000-kilometer instruction visit to troops in the field. He showed a twelve-minute, British-made video to the troops on the basic rules, keeping it more simple than his work with officers. The video deals with how troops should behave after taking a rebel house—how to treat civilians, rebel captives, and any Red Cross workers on the scene.

The third and equally important objective of the ICRC curriculum is to convince political, military, and police personnel that observing the laws of war strengthens their political position in relation to the rebels. Included are lessons on the negative consequences of violating international law—another reason to take law seriously. The existence of two UN war crimes tribunals, the reduction in the number of countries willing to harbor war criminals, the increased extradition of terrorists, and the cashiering of war criminals have all helped with this lesson. Of course, in order for ICRC delegates to maintain the principle of neutrality, this objective is never phrased as helping the army win the war, but the point could not be clearer. One ICRC publication presents the argument candidly.

> [M]ilitary leaders know very well that a murdering and plundering army is not worth much in military terms. In other words, respect for humanitarian rules is an element of discipline, which is an essential characteristic of an effective military unit. To put it simply, observance of international humanitarian law is not merely a burdensome duty, it is clearly in the interests of commanders of the armed forces.[35]

Another publication gives an additional reason for armies to obey international humanitarian law.

> While military operations that are conducted in violation of the law of war may appear to be successful in the short term, in the long run they are anything but. For example, all true military leaders know that unlawful acts strengthen the adversary's will to resist.[36]

I asked the delegates to state their opinion on the effectiveness of their dissemination efforts. They said it was hard to measure, but that it was their duty to try and that there were "some good trends" in finding more receptive people in the military. Although they were reluctant to judge their effectiveness, historical trends in Guatemala indicated that abuses steadily diminished until the date the final agreement was signed. There have been few abuses since. This means that more military officers are finding professionalization and legal conduct necessary to advance in rank. Military officers, like all bureaucrats, do what is necessary to advance. If they are told that massacring the civilian base of support for the guerrilla forces is necessary, some will do so to reach greater command. If that "necessity" is changed to observing recognized legal norms to generate more respect for the government, then ambitious military officers will change their modus operandi. The recent cashiering of Colonel Alpírez and others helped drive this point home.

Winding down the war was a team effort. The ICRC's position on the team was essential in accomplishing one conflict-resolution task: to get the military secure in its professionalization and willing to allow negotiations with the URNG. ICRC's dissemination efforts still contribute by enabling the civilian government to be more in harmony with the military when it carries out agreements on land reform, the return of displaced persons, and on human rights. In addition and equally important, the more the army follows the restrictions of Article 3 and Protocol II, the more confidence former URNG fighters will have in remaining alive. In effect, the government gets stronger, and the stronger it is, the less the interests of the military forces will be threatened by the peace accord. A key interest of the army is receiving amnesty for its war crimes, and the final agreement of December 29, 1996 is broadly forgiving on this issue. Some who had spoken out against amnesty before the final agreement had been murdered. A strong civilian government can make and enforce an amnesty policy, thereby giving protection and security to all in the military establishment except for a handful of particularly bad apples.

An observer can conclude, even if ICRC delegates cannot for political reasons, that the ICRC is doing the authorities a favor. Those governments that are unwilling to reform will invite popular revolutions. Reform, therefore, provides security to elites at a small price. In fact, why should civilized behavior be considered a price at all?

While the ICRC helps create the conditions that made the warring parties willing and able to negotiate, other players also help advance a civil society, setting the stage for third-party mediation.

The build-up of civil society is the complement to the "build-down" of the army's political role. The process is one of strengthening moderating organizations that exist between the unreconstructed political extremes of army and guerrillas. These organizations inject into politics players who have no interest in promoting the total victory of one warring party and the destruction of the other. These organizations promote law for their own protection and for building the civil society. They moderate the behavior of the army and guerrillas because, being in the middle, they can move away from any side that persists in indiscriminate violence and move toward any side that seeks a peaceful resolution to the conflict. In classical political terms, they become the "balancer" in the balance of power. The threat of an alliance with one wing keeps the other wing receptive to the interests of the moderating center. The center's political power is magnified if its members can vote among contenders for office. This is why all civil societies always have a strong political center. Without it, a polarized society, left and right, is organized to fight, not to contest elections in which the losing side would face destruction. Guatemala had experienced 35 years of polarization.

The organizations available to build the center in Guatemala include the Catholic Church; democratic political parties; the modern business sector; unions; intellectuals; and civic organizations representing professionals, Indians, women, and the landless. These groups have been growing in strength, especially since the return to ostensible civilian rule in 1986. In 1993, the attempted *autogolpe* by President Serrano with some army backing dramatically increased their political presence. Serrano's scheme to retreat to autocracy, reversing the trends towards a civil society, emboldened and linked these organizations.

The business sector responded to the *autogolpe* first, establishing a commission within its umbrella group, the Coordinating Committee of Agricultural, Commercial, Industrial, and Financial Associations (CACIF). Its ad hoc commission then organized a coalition of political parties, unions, and popular organizations to form the *Instancia Nacional de Consenso* (Committee of National Consensus). According to Rachel M. McCleary, a leading scholar of Guatemalan politics:

> The objectives of the *Instancia* were: to peacefully return the country to constitutional order; to promote the participation of different sectors of civil society in the decision-making process; to demonstrate to the international community the Guatemalan people's commitment

> to democracy; to strengthen democratic political institutions by restructuring, cleansing, and reforming those institutions; to consolidate the peace process; and to establish an intersectoral dialogue with the purpose of reaching consensus on the problems facing Guatemala and drawing up a national plan for the country.
>
> The *Instancia* was the first time in contemporary Guatemalan history that leaders from different parts of society who, under normal circumstances, would oppose each other ideologically and politically, came together to reach agreement on returning the country to institutional order through legal means.[37]

The selection by congress of a human-rights ombudsman, Ramero de León Carpio, as the interim president can largely be attributed to the work of the *Instancia.* The democratic momentum focused on the November 1995 general elections. Nobel Peace Prize winner and the champion of Guatemalan women, Rigoberta Menchú Tum, toured the country urging voter participation.[38] The URNG called a cease-fire for both rounds of the election (a presidential run-off occurred in January 1996) and urged people to vote.

Alvaro Arzú of the Party of National Advancement (PAN), the respected former mayor of Guatemala City, emerged as the new president in a closely contested election among two center-right candidates. In the legislature, six seats were won by a new party, the New Guatemala Democratic Front (FDNG), whose platform of land reform, military professionalization, Civil Patrol dissolution, and human rights reflected the agenda of the URNG. In effect, the guerrillas legitimized the constitutional order.

During this time, third-party mediation and support for democracy gained momentum as well. The UN, in particular, was a key member of the team determined to end the war. During this time as well, violence and human-rights violations continued until the election hiatus. "There is neither war nor peace," one observer remarked to me after the general election but before the cease-fire.

Until January 1994, mediation in Guatemala was conducted by Monsignor Rodolfo Quezada Turuno of the Guatemalan Catholic Bishops' Conference. A Framework Accord of that date created an Assembly of Civil Society (ASC), which brought together all segments of society (except some members of the agricultural elite) to resolve issues perpetuating the war. The Framework Accord also created the UN Verification Mission (MINUGUA) and a UN moderator position to facilitate a peace settlement.[39] The greater participation by representatives of the international community added more voices to the center. In support of the UN, a Group of Friends—consisting of Mexico, the United States, Norway, Spain, Colombia, and Venezuela—was formed to facilitate a settlement.[40] Since 1990, these governments hosted a series of meetings between guerrilla leaders and Guatemalan officials. Financial assistance for Guatemalan economic development, with pledges over a half billion dollars, came out of a 52-nation Consultative Group meeting in Paris in June 1995. It was designed to strengthen the peace process by assuaging the military, bolstering the modern business sector, supplying arguments on behalf of reform to be used by reformers, and giving hope to the poor.

The role of MINUGUA and UN moderator Jean Arnault has been widely praised. Established after the November 1994 partial peace agreement, MINUGUA has an annual budget of $20 million and a staff of 416 from 39 countries, including 50 police advisors.[41] As of early 1997, these UN officials were still awaiting the arrival of a new monitoring force to replace them, as called for in the final agreement. They were dispersed throughout the country to receive complaints about human rights abuses and investigate their validity. Between August and December 1995 alone, they received 3,000 complaints.[42] MINUGUA talks to all political factions. It issues quarterly reports, in which the army has been accused of such acts as harboring car-stealing and kidnapping gangs, murdering Indians, and employing forced conscription of young men off of buses and off the streets.[43] These reports have stopped forced conscription and, in the case of the Xamán massacres, provided the rationale for resignation of Defense Minister General Mario Enriquez. Above all, MINUGUA has created a climate in which the press can report corruption and in which more citizens can speak their minds with less fear of death at the hands of security forces. Protests by farm workers are increasingly allowed to occur without repression, as MINUGUA officials look on.[44] "We made it possible to say inside Guatemala what everyone was saying outside of Guatemala," commented the head of one of MINUGUA's regional offices.[45]

Jean Arnault, the UN moderator, through negotiations in Mexico City and elsewhere, brokered the 1996 cease-fire and final accord.

The network of third parties, both internal and external, methodically created the conditions that ended the civil war—conditions in perfect disharmony with the conditions that began and perpetuated the war. The ICRC played a pivotal role, fully in keeping with its legal mandate according to the Geneva Conventions.

This case is a textbook example of how forces to end war and violence can engage in a multiple assault. It is hard to leave Guatemala without a sense of optimism.

This optimism is still tempered by continuing army abuses. The military remains involved in drug trafficking, thefts, and protection rackets. One businessperson told me in a hushed voice that if his business got any larger the army would demand payment for his staying in business and alive. He remains skeptical. Yet, the final peace agreement calls for the military to cease all police functions and to reduce by one-third its personnel—now more professional thanks to the ICRC. This supports a sense of optimism for peace.[46]

But even this optimism is tempered by some grim socioeconomic demographics. Infant mortality rates remain high in the rural areas. Even though the growth rate of the gross domestic product has reached a respectable 4 percent, its distribution continues to be skewed to the wealthy. The underemployment and unemployment rate of 48 percent has not changed in twenty years.[47] Worst of all, with a population growth rate of 2.8 percent, a rate at which Guatemala's population will double every 26 years, the country cannot provide education, sanitation, water, housing, health care, and employment for all. Guatemala City, the largest urban area in Central America, sprawls, its shanties seep into ridges and valleys like an uncontrolled oil slick. These are not conditions for domestic harmony.

Concerted efforts at socioeconomic development, the ICRC rightly believes, must become the first line of defense against more nasty internal wars that would pit rich against poor, tribe against tribe, religion against religion, and ethnic group against ethnic group. A second line of defense is the international community's efforts at preventive diplomacy. Neither lines of defense can be manned without active participation by and coordination of governments, IGOs, and NGOs. The ICRC is increasingly engaged in publicizing the necessity of both lines of defense to reduce the scourge of war. Perhaps encouraging socioeconomic development and preventive diplomacy could be called its third mission.

A description of the ICRC's efforts on behalf of socioeconomic development and preventive diplomacy will be given in the following chapter.

7

A New Contract of Humanity

ICRC President Cornelio Sommaruga began his keynote address to the 26th International Conference of the Red Cross and Red Crescent by saying: "The world is weighed down by the victims of too many tragedies."[1]

The December 1995 conference, as previously analyzed, primarily sought to mobilize the entire Movement and the international community toward moderating and ending today's tragedies, those brutal internal wars in which international humanitarian law is violated daily by the warring parties. Two other tasks were given prominence at the conference. They round out the ICRC's expanded agenda on wars. One is "preventive action," and the other socioeconomic development. The two tasks are linked. With more preventive action or preventive diplomacy, fewer wars will break out, and therefore greater resources can be devoted to development rather than to arms and killing. And, with more development, the ICRC believes, the fewer socioeconomic motives for wars such as class or ethnic dominance will be present. President Sommaruga calls the entire package "a new contract of humanity."[2]

This Wilsonian-sounding phrase explicitly reflects the fact that the post–Cold War era of internal wars has imposed upon the ICRC a "shift towards a more 'system wide' approach" in its activities.[3] This shift is explained in an ICRC report entitled *Challenges of the Nineties,* given to those attending the conference. President Sommaruga wrote the preface, "The Shape of Things to Come," in which, after describing the coming of the tragic internal wars, he outlines the system-wide agenda.

> The ICRC has tried to rise to the many challenges these developments [wars] have posed. The 1990s have marked the "coming out" of the organization in many senses: not just in obtaining observer status at the United Nations General Assembly, but in its willingness to cooperate with other humanitarian players in the interest of the victims, with United Nations agencies, multilateral donors and non-governmental groups; to collaborate more closely with human rights and other specialist lawyers as the relevance to them of international humanitarian law grows; to be more public about its concerns—whether regarding the security of civilian populations or, for example, on the health effects of

> weapons such as anti-personnel landmines; and its willingness to broaden its activities to help enable populations and victims not just to meet their immediate relief needs but also to prepare themselves to recover their productive lives. Above all, the ICRC has intensified its collaboration with other components of the Red Cross and Red Crescent Movement, both with National Societies and the International Federation of Red Cross and Red Crescent Societies, recognizing the unique capacity of the Movement to deliver services to the victims. This has been evident, for example, in tracing, dissemination and public information activities.
>
> The ICRC remains braced for the future, though not without anxiety. Global socio-economic currents threaten to put ever greater strains on populations around the world. The battle for resources will continue to be compounded by deep-flowing political and ethnic tensions. Governments in many countries will find it increasingly difficult to respond to the aspirations and basic needs of their own people, and there will always be the unscrupulous willing to take advantage of wider disaffection for personal gain. Debt burdens, unregulated transfer of weapons and enormous pressure on natural resources will take a greater toll on the most vulnerable and discriminated against in many societies.[4]

The ICRC is indeed "coming out." If this new activism described by President Sommaruga had occurred during the Cold War, it would have been seen as taking the side of the West, and thus avoided. Now, there are no sides. All states are ostensibly committed to advance human rights. A sense of "anxiety" can also be recognized as the premier war relief agency takes on additional tasks, tasks that go beyond the letter but not the spirit of the Geneva Conventions and Protocols. Politics by any other name—or no name at all—is still politics. Conflict prevention and resolution will always be the "stuff" of politics, and the ICRC is increasingly playing that game. However, as President Sommaruga explains, the ICRC "has tried to meet the challenges" of the post–Cold War world. Well, why not? If our initial analysis is correct—that the likelihood of international wars has decreased to the point that a long peace now embraces many still smoldering regional disputes—then is it not true that internal wars are now the only games in town? These are today's wars. These are where war victims are. These wars are where the ICRC is needed.

These are Article 3 wars.

Once involved, the ICRC carries out its main mission of humanitarian relief and protection of war victims, engages in its unspoken mission of making these wars dysfunctional so as to end them, and, increasingly and more modestly, begins to help in conflict prevention and socioeconomic development.

CONFLICT RESOLUTION AND SOCIOECONOMIC DEVELOPMENT

It should not surprise the reader that the ICRC's strategy for these two interrelated tasks heavily involves organizing international meetings, building consensus on new norms at these meetings, and issuing reports—its "dissemination and public information activities"—in order to guide the international community. This strategy is standard operating procedure for ICRC lobbying.

The strategy targets specific groups, namely: National Societies and their International Federation, United Nations agencies, governments (especially their military and legal establishments), NGOs, and the media. The reason for this emphasis on other organizations is clear. Unlike the ICRC's main mission or even, to a lesser extent, its unspoken mission, promoting preventive action and socioeconomic development depends overwhelmingly on the actions of others. The ICRC does have a few important operational roles, but not many. It must work through other organizations. I am tempted to call prevention and development the ICRC's "spoken mission," because ICRC leaders mainly talk to other groups and individuals, urging them to action.

National Societies and Their International Federation

The recent flood of internal wars elevated the importance of National Societies and their International Federation as partners in ICRC's missions. In international wars, the National Societies work as auxiliaries to the military medical services of their home countries. Some distance between the ICRC and National Societies is therefore necessary in order to uphold the ICRC's neutrality, impartiality, and independence. In internal wars, national societies surround the entire war, providing humanitarian assistance to everyone in need. This universal access and the National Societies' resources and familiarity with the local culture make them natural partners for the ICRC.* In the Chiapas conflict in Mexico, for example, the ICRC worked with the Mexican Red Cross in creating "neutral areas" for negotiations, in disseminating international humanitarian law to both sides, in evacuating the wounded and civilians from "danger areas," and in assisting displaced people.[5] The ICRC can modestly claim that its action in tandem with the Mexican Red Cross avoided the resumption of hostile action in many dicey instances between the government and Zapatista rebels.

---

*Governments fighting rebels may not like National Societies to have universal access, and they often try to restrict their fields of operation.

While stopping armed conflict is important, preventing conflict is obviously more effective in protecting civilians. Here, too, the National Societies have a role to play. In 1994, the leadership of the ICRC began a campaign to convince the National Societies "to help the most vulnerable groups" in their countries.[6] Yves Sandoz, director of Principles, Law and Relations with the Movement, urged that National Society efforts "must be made to eliminate the root causes of armed conflict," among which are: "poverty, inequality, under-development, over-population, illiteracy, and crime."[7] National Societies must "preach humanity."[8] At the 26th International Conference, President Sommaruga urged the National Societies "to promote the value of tolerance."[9]

Working with the most vulnerable groups in countries susceptible to internal war and taking up their causes further thrusts the Red Cross and Red Crescent Movement into politics. Vulnerable groups do not control governments. They are vulnerable, in fact, because governments neglect or repress them. To redress these inequities, political action is required.

In response, National Societies are beginning to come to a consensus that they should play greater political roles at home. The president of the International Federation of Red Cross and Red Crescent Societies, Mario Villarroel Lander, acknowledged at the International Conference that

> [T]here is an enormous force in the world today directed towards caring for the most vulnerable. It is because this force exists and works that I can speak here of hopes and be sure that they will be fulfilled.[10]

National Societies avoided such a role in the past, the Federation President admitted, "fearing it to be too political."

> Yet, who else but ourselves is in a better position to observe and assess the adverse effects of ethnic strife, discrimination against minorities, natural disasters and UN sanctions that inadvertently affect innocent people?
>
> Was Henry Dunant too political in advocating the lessons to be drawn from his experience at Solferino before all the courts of Europe? Was Henry Davison too political in seeking the support of governments for the action of the League of Red Cross Societies in devastated Europe after the First World War? We need to follow their examples with courage and do our best to develop humanitarian action in a sensitive and discriminating way and exercise our influence to make sure that it is implemented. By this I mean that the Red Cross and Red Crescent should be the spokesman of the world's most vulnerable, a voice that is heeded by all peoples and states.[11]

The rhetorical nature of his questions indicates that moving the Movement into more active and overt, albeit humanitarian, political roles is only beginning. Conference papers can proclaim that "relief is much more than providing tents and blankets," but preventing internal wars is a huge political task.[12] To speak for societies' most vulnerable is to raise their political status and interests.

The related task of socioeconomic development is less political, but only slightly less. Programs that develop food self-sufficiency; maintain village social structures; and create mechanisms for giving the poor, ill, and abused a political voice capable of positive government responses will affect relations between the rulers and the ruled. For these tasks, the Movement needs outside help and security. The imprimatur and assistance of UN agencies is key.

The United Nations

The ICRC knows that the international community, especially the UN, needs to lead the way in preventing conflicts. We have already described the ICRC's lobbying efforts to get the necessary political and military clout injected into internal wars in order to help make them dysfunctional and to end violence.

When internal disturbances that can lead to war occur, the ICRC is fully prepared to take an integrated approach with political, military, and other humanitarian agencies to prevent war. The ICRC is fully in accord with former secretary-general Boutros-Ghali's *An Agenda for Peace* in the prewar phase of internal disturbances. Boutros-Ghali advocated concerted efforts by all third parties to reconcile potential warring parties. For example, after the bloodletting in Burundi in October 1993, the ICRC worked with the secretary-general, UN monitors and mediators, key governments, and NGOs to reconcile the Tutsi and Hutu. Ancient hatreds fueled clashes between the minority Tutsi-dominated army and the majority Hutu-dominated government and militias. The ICRC, for its part, mediated with both tribes, helping them to draft "a set of rules of humanitarian behaviour that are particularly applicable in times of internal violence, and to working out suitable strategies for making them known."[13] Using these agreed-upon rules as political protection (these are your rules—we're impartial and neutral), ICRC delegates in the field provided "humanitarian guidance to authorities and communities at the first signs of tension."[14] As always it was important to avoid being accused of taking sides, and getting both sides to follow rules they have already agreed upon can be explained as politically neutral behavior. Thus, the ICRC broadcasts messages on Burundi radio and television, according to ICRC spokesman Tony Burgener, explaining his organization's neutrality and humanitarian goals.[15]

Needless to say, the UN-led network has been active in Burundi since 1993 because signs of tension are a daily phenomenon. Bloody massacres are all too frequent, and many thousands have been killed, including three ICRC delegates. But for three years, the third-party presence has inhibited a full-scale slaughter like that in neighboring Rwanda in 1994. The ICRC fully supports the current (as of 1996) mediation by former Tanzanian president Julius Nyerere and gives support behind the scenes. "This is the right approach, I feel," said President Sommaruga in explaining the general policy at the International Conference, "for *conflict prevention.*" He continued:

> [P]reventive diplomacy, economic aid, development aid, humanitarian assistance and the deployment of military observers can contribute significantly to stabilizing situations which might otherwise degenerate. The same type of synergy can come into play in the *post-conflict* phase, when peace must be consolidated, reconstruction work begun and, in many cases, humanitarian activities conducted for the most needy.[16]

ICRC policy changes during the "acute phase of a conflict," during which prevention has obviously failed and socioeconomic development is impossible. An integrated approach with UN peace enforcers, as *An Agenda for Peace* favors, would link the ICRC with political/military action, and that would destroy "the neutral and impartial nature of humanitarian action."[17] Access to victims with the consent of all the warring parties becomes the top priority during a war. According to the Geneva Conventions, "there must be room for independent humanitarian action" to assist and protect war victims.[18] During wars, the ICRC wants the UN to be fully involved in political-military work while ICRC workers preserve their autonomy so they can provide relief and protect civilians and prisoners on all sides. This arrangement maximizes the ICRC's main mission *and* its unspoken mission of undermining and resolving internal wars.

Aware that preventive diplomacy and UN socioeconomic development efforts depend on the support of UN members, the ICRC seeks to make governments provide that support.

Governments and Their Experts

The ICRC targets governments and to a lesser extent, dissident warring parties, in disseminating international humanitarian law. Dissemination is a constant for the ICRC. While most of international humanitarian law prescribes how warring parties must act toward prisoners, the wounded, civilians, cultural objects, and so on, the ICRC recognizes that for its unspoken mission of making wars dysfunctional and ending them, as well as promoting preventive diplomacy, getting governments *to enforce international law*—both as governments and as UN members—is essential. If compliance with international humanitarian law could be increased, then *internal wars would be extremely difficult to conduct.* That is, if war victims could be protected, how could they be murdered, terrorized, or forced to evacuate territory? How could their political will be broken? How could the warring parties be victorious if they could not wipe out the enemy's civilian base of support? For preventive diplomacy, the ICRC's thinking goes one step further. If the enforcement of international law became institutionalized and programmatic, then potential warring parties would be deterred from going to war! Not only would their war crimes induce foreign intervention, but they would be made accountable for their criminal behavior.

So, to increase law enforcement, the ICRC puts into effect its standard strategy. First, call a conference. In 1992-93, the Swiss Federal Council was lobbied to call an International Conference for the Protection of War Victims. It met in Geneva, from August 30 to September 1, 1993. The states signatory to the Geneva Conventions were invited. Not surprisingly, the Final Declaration condemned grave violations of law.

> We reaffirm the necessity to make the implementation of international humanitarian law more effective. In this spirit, we call upon the Swiss Government to convene an open-ended intergovernmental group of experts to study practical means of promoting full respect for and compliance with that law, and to prepare a report for submission to the States and to the next session of the International Conference of the Red Cross and Red Crescent.[19]

After a preparatory meeting in 1994, experts representing 107 states and 28 governmental and non-governmental organizations met in Geneva, January 23-27, 1995. They prepared a set of recommendations that became a substantial part of Commission I's deliberations at the International Conference. We will not review those that pertain to limits on the conduct of war—most of which were discussed in Chapters 2 and 4, especially those that provide for protection of women and children. Instead, we will concentrate here on those that would help prevent or deter war.

At the meeting of experts, the ICRC proposed that time not be wasted in an overall revision of international humanitarian law. Such a revision could take over 30 years. Rather, clarifying and applying existing law should be the prime agenda. Dissemination was stressed, including ways of "persuading dissident forces to comply with IHL."[20]

Two measures the ICRC recommended to the government experts for dealing with war crimes also would have a preventive or deterrent effect on war.

One was the use of the International Fact-Finding Commission established under Article 90 of the 1977 Additional Protocol I. This would investigate violations of international humanitarian law in both international and internal wars. The other was the institutionalized punishment of war criminals in state courts and the establishment of an international criminal court.[21] In their final report, the experts recommended that states do exactly what the ICRC had recommended.[22] If war criminals are exposed and punished, the long-term effect would likely be a reduction in war crimes and, given the nature of internal war, a reduction of those wars. Potential warlords might hesitate to order combat that would violate the laws of war and make themselves into war criminals. Resolution 1 of the International Conference of Red Cross and Red Crescent endorsed both the 1993 Final Declaration of the International Conference for the Protection of War Victims and the 1995 Recommendations drawn up by the Intergovernmental Group of Experts.[23]

On the issue of getting governments to increase socioeconomic development aid, the ICRC has been less successful. Like most international humanitarian agencies, the ICRC is disappointed that aid by the major states is sluggish at best. Over 1 billion people subsist on less than a dollar a day.[24] During the Cold War, foreign aid was the superpowers' payoff for allied loyalty or for preserving Third World non-alignment. Resources flowed to less-developed states. Now, Russian foreign aid, besides being a fraction of what it was, is confined to its "near abroad," a few former republics of the old Soviet Union. The United States, too, is cutting its USAID budget and the number of states it assists. In absolute dollars, Washington's budget is now behind those of Japan, France, and Germany. For 1995, U.S. aid to developing countries dropped 26 percent, solidifying its last place position in giving aid as a percent of GDP among the 21-nation Development Assistance Committee within the Organization for Economic Cooperation and Development (OECD). The OECD reported that overall aid to the developing world fell from $59.2 billion in 1994 to $59 billion in 1995.[25]

Every chance they get, ICRC leaders at the UN, at international conferences, and in their publications call for states to be more active and coordinated in providing developmental assistance. "[I]t is worth remembering," President Sommaruga reminded the governments attending the 26th International Conference, "that the chasm between the developed world and an ever-growing proportion of the planet's population is widening daily. Unless we take care, this glaring disparity will be the cause of tomorrow's conflicts."[26]

Even where the guns of internal wars have been silenced by cease-fires or peace accords, the ICRC believes that the failure to correct "the problems that have led to a people's vulnerability" will lead to "a recurring and evolving

Providing Water in Nepal.

The more the civilian population is self-sufficient, the less it must depend on warlords. Working with National Societies and their International Federation, the ICRC in real and potential war zones advances agricultural production, animal husbandry, water supplies, and sanitation. This work lessens the motives for rebellion by satisfying the people and fostering their independence. While installing a water system, for example, Red Cross workers train people how to do it themselves in the future. *(Photo courtesy of International Federation, by Liliane Toledo)* cycle" of conflict.[27] Today's internal wars have a nasty habit of rekindling: Afghanistan, Somalia, Liberia, Sri Lanka, Angola, and Sudan have all experienced resumed hostilities. People need the hope that socioeconomic progress will come to them. Reconciliation is cemented when the "haves" take an interest in and mobilize the "have nots" toward a better life. That is progress. Without a plan for that progress, feelings of exploitation, repression, and alienation will linger and only await a rekindling spark. The leading governments in the international community can provide local elites the resources to give "the poor, ill and abused" a full voice in the community in order to "solve today's mounting problems."[28]

While still relying on governments and the UN to carry the burden of socioeconomic development in states vulnerable to domestic strife, the ICRC has increased its grassroots efforts by providing agricultural tools, fishing tackle, seed, water pumps, and veterinary assistance to war-ravaged areas. In all cases, ICRC workers in the field work through local leaders, concentrate on training, and seek to conserve the environment. They also try to get NGOs to emulate their programs.

Non-Governmental Organizations

The ICRC's big complaint about NGOs in the field is that some engage in high-profile, uncoordinated relief operations during media-covered disasters, and then cut and run when the disaster seems stabilized and the media leave. The ICRC wants NGOs to coordinate with the Red Cross and Red Crescent Movement during crises and pay attention to rehabilitation work.[29] Once the guns stop, the country in which they were used is always at the height of devastation. The full measure of destruction has been delivered. That is when the need for rehabilitation is the greatest and when hope must be nurtured. Although it is hard to measure, most established NGOs now think in terms of the long-haul, heeding the admonitions of the ICRC. The media is also getting this message.

The Media

The ICRC's films, audio-visuals, CD-ROMs, street theater, commentary pieces in the world's major newspapers, journal articles, and radio and television broadcasts have all been increased in number with the growth of the ICRC's role in war. Internal wars are peoples' wars. Therefore, to moderate and end them requires that the dissident fighters as well as government armed forces follow international humanitarian law. The people who take up arms against their governments are not trained in the laws of war. They have been perpetrators of war crimes as well as victims. To get the people acquainted with humanitarian principles, according to the ICRC's Yves Sandoz, "the media are an essential link in this process. During times of conflict, the media are often the only means of getting the humanitarian message through to the people."[30] Community leaders and young people are the principal target audiences. The message is a set of simple, basic rules on how to deal with tension and violence. The aim is "not only to prevent the escalation of hostilities and open the door to reconciliation but also to prevent armed conflict itself."[31]

For example, the ICRC has been touring a play in Burundi "illustrating the basic standards of humanitarian behavior that must be maintained in times of turmoil."[32] Over 25,000 people attended the show in 1995 alone.

If warring parties can preach and teach ethnic and religious hatred, then preaching and teaching the humanity of mutual caring is the antidote. Is there any historical evidence that a group has gone to war against another group it perceives to be caring for everyone's needs and wants? People shoot enemies, not friends—at least in politics.

CONCLUSION

No one expects the motives for war to disappear overnight. Class exploitation, ethnic and national separatism, religious attempts to "purify" cultures or seize power, and territorial expansion continue to feed regional and internal conflicts. Because of this, the ICRC focuses its unspoken mission on the resolution of these remaining conflicts. It wants wars to be a less attractive option than peaceful negotiations, and the early returns show that its efforts as a player in the international community are working. Wars are becoming increasingly dysfunctional. The Stockholm International Peace Research Institute reports that the number of major armed conflicts has declined since 1989. Even with new wars breaking out, the overall number dropped from 32 in 1994 to 30 in 1995.[33]

This trend indicates that other means of conflict resolution will be used. In disputes within states, protests, civil disobedience, political party organizing, and the development of a civic culture are increasingly favored strategies to put divided societies on the road to democracy. The brave leadership of Aung San Suu Kyi in Burma provides a good example. Internationally, third parties use economic sanctions, diplomatic isolation, and diplomatic/military intervention to thwart war and open the option of negotiation.

Still, current internal wars continue and new ones break out in spite of the warring parties' awareness that foreign intervention will be hard to avoid. Today's warlords must know that players and resources antithetical to war will inevitably be injected into the fray. The presence of the ICRC, in particular, is corrosive to the institution of war.

Won't this make ICRC delegates *persona non grata*? This question has nagged me since I began researching the ICRC's position towards war, and it still nags. As more warring parties become aware of ICRC efforts to push and pull the international community into moderating, undermining, and eventually negotiating settlements of their wars, won't the warring parties who think they can otherwise win try to restrict or prohibit ICRC access to the victims of war? Won't they keep out the people who want to protect and sustain the very targets of their combat?

Yes, but. Yes, the strong in war always try to isolate the conflict to preserve the power relations that have them on top. Deny outside aid to a weaker enemy and victory is assured. This is one reason the Russians initially allowed only a modest role for the ICRC, OSCE, or anyone else in Chechnya. Yes, warring parties, especially governments, are aware that when the ICRC protects civilians and makes them more self-sufficient; disseminates international humanitarian law; and expands pro-peace, third-party players, it inhibits their combat. Naturally, they will try to manage humanitarian efforts that would dampen their war. But keep them out completely? Not likely. Why?

The ICRC knows that it has the imprimatur of international law behind it. This is solid ground. It has a right under the Geneva Conventions that 188 states have signed to provide humanitarian assistance in war. Governments that are parties to the Geneva Conventions are obliged to accept its humanitarian efforts. Even dissident groups, such as the Zapatistas in Chiapas, recognize the legitimate services the ICRC can perform in their struggle.

The ICRC knows that the UN, its peacekeepers and war crimes tribunals, look unfavorably on warring parties who perpetrate barbarities and prevent ICRC access or kill its workers. The international community is on the side of the ICRC.

The ICRC knows that virtually all warring parties need the services it provides. Who else is equipped to visit prisoners and detainees? Who else can trace displaced persons? Who else can mobilize National Societies from around the world to aid in medical programs, food relief, and other humanitarian assistance?

The ICRC knows that in acute phases of war it can open room for its humanitarian action by distancing its operations from political and military third parties. When war rages, the ICRC can be neutral, impartial, and independent, and thus it can appear to not affect the military equation. It can reduce the jeopardy of its workers in the field. It can count on the political/military third parties to handle the negotiations to end the war.

The ICRC knows it is in for the long haul. It sees progress in the expansion and enforcement of international humanitarian law, from banning antipersonnel mines to bringing war criminals to justice. As its founders Henry

Dunant and Gustave Moynier anticipated, it sees the institution of war ever constricted. Even in the most disgusting wars, its delegates are willing to takes risks to stop the barbarity.[34]

ICRC strategies will be tailored to each particular "state of the war." It may require sustaining and protecting the civilian population so it can endure, as in Sudan or Afghanistan. Eventually, war weariness and frustration will induce the warring parties to settle. It may require working with and encouraging UN-mandated peacekeepers, as in the former Yugoslavia. Eventually, with victory impossible, conditions will be made right for a peace treaty. It may entail developing close relations with political and military authorities, as in Guatemala. Eventually, government officials will treat their people to enough human rights to strip away the motives for rebellion and prepare the way for third parties to help formalize democratic methods to manage political conflicts.

There will be no retreat for this proud, strongly self-directed, global, professional, and modestly self-righteous organization that is singularly buttressed by international law.

There will be no retreat from trying to prevent wars, giving humanitarian aid and protection to victims if war breaks out, organizing third-party efforts to moderate and settle the wars, disseminating international humanitarian law, and encouraging the international community to provide rehabilitation aid so wars do not rekindle. This is the essence of the ICRC's new contract for humanity. The hideous institution called war is in trouble.

8

Epilogue: Chechnya and the Hazards of the Unspoken Mission

Six ICRC delegates were murdered in Chechnya while this book was in production. This barbaric event again highlights the hazards ICRC delegates face daily in the field; hazards not sufficiently analyzed perhaps in previous chapters. Hence, this epilogue.

When the ICRC's unspoken mission becomes generally known, new questions will arise about how to minimize the dangers of humanitarian action in light of its corrosive effects on internal wars. The hazards need to be identified first and in greater detail. Then, ways to minimize them can be formulated. The Chechen case is as good as any to use as the backdrop for this analysis.

ANOTHER WAR, ANOTHER CRIME

It is difficult to imagine Chechnya as a happy place anytime in its history. Isolated in the North Caucasus mountains—but not isolated enough—it was an economic backwater on the fringe of the Ottoman Empire. Intensely clannish, it maintained feudal and feuding traditions. The culture was xenophobic. And for good reason. Foreigners kept fighting for control of the region—the Ottomans to preserve a barrier to the expanding Russian empire, and the Russians to break down that barrier. The Chechens kept fighting for independence.

In 1859, after five decades of scorched-earth warfare, the Russians emerged victorious over the Chechens and other area tribes. More than once, violent opposition to Russian occupation more than once brought the destruction of a village in retribution.

When the Russian empire fell apart after the 1917 Revolution and the subsequent civil war, the Chechens made another attempt at independence. The attempt failed, and it only added to Chechnya's reputation as a land of wild, uncivilized "mountaineers," a reputation still accepted today by most Russians. In 1944, Stalin deported Chechens en masse to Kazakstan and Siberia on the charge that they were collaborating with the Germans. One third of the deportees reportedly perished in the process. The others returned, embittered, during the "Thaw" under Khrushchev in the 1950s.

In 1991, when the Soviet Union began disintegrating, the All-National Congress of the Chechen People (ANCC) proclaimed independence under the leadership of a former Soviet air force general, Dzhokher M. Dudayev. In October 1991, Dudayev was elected president. This suddenly and belatedly gained the attention of Russian President Boris Yeltsin. He decreed emergency rule the following month. In January 1993, a deal worked out by Russian parliamentarians with Chechen parliamentarians provided for only a federation. Dudayev was excluded then and also during a follow-up conference in March 1994. But Dudayev's supporters controlled the capital, Grozny, and most of the central and southern part of the country. Determined to crush the separatist movement, Yeltsin ordered that a pro-Moscow faction in the north be armed and supported in an attack on Grozny. After this failed, and after being convinced by his defense minister Pavel Grachev that Russian forces could take Grozny in hours, Yeltsin ordered an attack set for December 11, 1994. Grachev's hours became weeks then months for 30,000 Russian Federation troops. The Russian air force bombed Grozny into rubble and made villages disappear. After cease-fires, the taking of hostages, and more bombing, the Russians eventually occupied Grozny and established a puppet government in March 1995. But even Dudayev's death in April 1996, the result of a rocket attack, did not halt the rebellion. The rebels had friends abroad who supplied dollars, communication equipment, and small arms.

Russian national interests in keeping Chechnya in the Federation are well known. Once one constituent republic separates, where would the disintegration stop? Chechnya also has major oil and gas reserves and is traversed by an important pipeline. Finally, Russian nationals make up about 20 percent of the population.

Nevertheless, the failure to end the war played badly in Moscow and throughout the Federation. Yeltsin, therefore, moved to remove the war as an issue in the upcoming June 1996 presidential elections. On June 1, a cease-fire was negotiated, and after the first round of the elections, Yeltsin made the losing third-place candidate, Alexandr I. Lebed, his national security advisor and charged him with ending the war. This made Yeltsin the favorite of the ANCC. Fighting stopped, but segments of the Russian military violated the terms of the cease-fire by not dismantling their checkpoints.

On August 6, 1996, Chechen forces assaulted Grozny, defeated the Russians, surrounded the 3,000 or so of its troops still remaining in Grozny, and demanded an agreement on the question of independence. The Chechens got something close. Lebed concluded an agreement with the Chechen military commander, Aslan Maskhadov, on August 31, that called for a joint commission to manage law and order, troop withdrawals, and reconstruction. The question of independence would be settled within five years. Yeltsin gave the agreement his wavering support, finally firing the ambitious and immodest Lebed in October.

Even after the agreement and cease-fire, Chechnya remained a dangerous place. While warfare between forces of the Russian Federation and the Chechen separatists no longer produced all the horrors of civil war—already with a toll of 40,000 dead (90 percent civilian) and about 400,000 displaced—a continuing internal struggle between Chechen factions produced horrors of its own. Once it became clear that the Russians might be leaving, perhaps for good, the issue for the Chechen factions was one of power. Who would rule? Would it be the "moderates" who made the deal with Lebed or a more radical faction that included Islamic extremists?

Leaders of the moderate faction hastily called for presidential and parliamentary elections to be held on January 27, 1997. They wanted a quick confirmation of their authority, ascendant after their military victory over the Russians and the subsequent cease-fire.

In an intensely clannish society, that had champions abroad, was infused with calls for Islamic purity, and was without established procedures to select leaders, most of the ingredients for internal strife were assembled.

ENTER THE ICRC TO HAZARDOUS DUTY

From the very beginning of the war, the ICRC and other international agencies had sought to bring humanitarian relief and protection, the resolution of the conflict, and, eventually, social and political reconstruction.

It has been a hazardous duty.

During the civil war, Russian aid workers were kidnapped and murdered. More than a score of journalists suffered the same fate. On a mission sponsored by the Soros Foundation, world-famous humanitarian entrepreneur Fred Cuny disappeared along with his translator and two doctors. Common wisdom accused the Russian security service of leaking the misinformation that Cuny was a Russian agent gathering information on the rebels. This led to his and his companions' murder by a faction of the Chechen separatists. The Russians, still pursuing their scorched-earth strategy, thereby eliminated a credible witness to their atrocities, a witness who could command widespread media attention.

Even after the cease-fire in August 1996, two election monitors from the Organization of Security and Cooperation in Europe (OSCE) were kidnapped (and later released unharmed). Twenty-one Russian police officers, authorized by both sides to conduct joint patrols with Chechen forces, were held hostage by a rogue Chechen field commander. They, too, were released after tense negotiations.

Those who conduct wars, civil or otherwise, do not like witnesses. Wars are brutal, and selective reporting can make one side or faction seem especially brutal, thus damaging its effort to build domestic and foreign support.

The ICRC serves as a witness. It also protects and provides humanitarian assistance and has tried to mute the Chechen war since its inception in 1994.

One of the ICRC's main facilities is its hospital at Novye Atagi, about 25 kilometers south of Grozny. Financed by the Norwegian Government and Norwegian Red Cross, it opened just days after the August agreement. The ICRC received the consent of both sides to open the facility and to provide medical services to all in need. Still, its 20 delegates and about 150 local staff were subject to harassment and violence. The motives of those who directed their wrath at the ICRC were sometimes trivial, sometimes not.

In early November 1996, an ICRC delegate was kidnapped from the fenced-in compound by a leader of a local Chechen faction. The leader was upset that another faction received the ICRC contract to supply drivers and hospital workers. With the local economy in shambles, those 150 or so well-paying jobs were prized. The settlement worked out by the ICRC reflected its principle of neutrality. The head delegate negotiated with the military superior of the kidnappers, re-distributed some jobs, and secured the release of the ICRC employee.[1] As with virtually all warring parties described in this study, there are many factions within each side. This makes war complex and inefficient. It also makes it and other political strife associated with war harder to resolve.

The kidnapping illustrates a number of the ICRC's difficulties in performing both its main and unspoken missions. In internal wars and power struggles, the leaders of political groups have to provide for their fighters and supporters. Foreign third parties, whether humanitarian workers or peacekeepers, bring goods and jobs to areas where the economy is disrupted and depressed by strife. They also bring essential services to sustain life—restoring water supply systems and providing medical assistance, in particular. These goods, jobs, and services are coveted, because to control them is to sustain one's group and to gain power over those factions who do not have such control or access. This makes the ICRC or any agent of humanitarian assistance a prize to be fought over.

The contest often gets rough.

The ICRC's principles of neutrality (not taking sides), impartiality (serve all victims according to needs), and independence (no subordination to political authority) command that the ICRC resist attempts by a contesting faction to co-opt its services or to deny these services to the opposition. This resistance has to be non-violent and non-partisan. To resist otherwise—and this is a catch-22—would itself violate the three principles. It would appear to the target of such resistance that the ICRC was on the other side and so hardly impartial, neutral, or independent. For protection in conflict areas, the ICRC depends primarily upon international peacekeepers and local authorities.

As a result, ICRC delegates must remain patient, discreet, non-violent, committed to negotiations, and willing to live with dilemmas.

I will digress briefly to emphasize these dilemmas. The reemergence of charges in late 1996 that ICRC delegates knew about and visited Nazi death camps and yet did not publicize their existence world wide graphically illustrate the nature of the ICRC's dilemmas. Blow the whistle on the Nazi brutality and the Nazis might remove the thousands of Red Cross workers who were providing life-saving services to millions of POWS and detainees inside the Third Reich. While seemingly indifferent to the holocaust, these services saved hundreds of thousands, if not millions of lives. During World War II, ICRC delegates made 11,170 visits to the camps, mostly in Germany, and had arranged for the delivery and distribution of 470,000 tons of relief supplies to POWs and civilian detainees. During the wartime famine in occupied Greece, over 750,000 tons of food and other supplies were distributed by the ICRC and the Swedish Government. Keeping quiet preserved access, maintained the principle of neutrality, and saved what lives could be saved. Other dilemmas have appeared in previous pages, such as one that arose over helping Bosnian, Croat, and Serb civilians evacuate areas targeted for ethnic cleansing, and therefore appearing to abet that heinous practice. The other distasteful choice was to leave the civilians in their homes and available for "cleansing." ICRC delegates are acutely aware that their missions create dilemmas; they express a resigned sadness and helplessness when discussing these dilemmas and a controlled anger when charged with the equivalent of war crimes. To claim they are callous to the fate of Jews, Muslims, or anyone else is to judge them unfairly. Their business is to save people, even at great risk to themselves.

The hazards in Chechnya—to return to the case study—continued.

In late November, just two weeks after the kidnapping, armed Chechen gunmen entered the hospital compound without being seen. One group began stealing medicine from a store room, and another began removing the ICRC's expensive communication equipment. The Swiss head delegate of the hospital, Christophe Hench, walked in on the robbery, failed to persuade the gunmen to leave, and helplessly watched as they stole what they wanted. "And they left a note in the radio room to the effect that we were spies and hypocrites," recalled Fery Aalame, head of the ICRC delegation in the North Caucasus.[2]

Much worse soon followed.

A few minutes before 4:00 A.M. on December 17, 1996, masked gunmen broke into the hospital compound at Novye Atagi. The two local guards, hired by the ICRC after the previous month's robbery, failed to detect the intruders. The gunmen entered the barracks and, using silencers, shot seven ICRC delegates in their beds. At that point, a guard became aware that the security of the compound had been breached and fired his AK-47 into the air. Although the guard had disobeyed ICRC orders forbidding him to be armed, the guard's warning undoubtedly saved lives. The gunmen fled.[3]

Of the seven delegates who had been shot, only one survived. Five of the six killed had been seconded to the ICRC by National Societies. The six who died were:

> **HANS ELKERBOUT,** 47, the construction manager, was from The Netherlands Red Cross. Elkerbout previously served as a civil engineer in Turkey, Afghanistan, Pakistan, Iran, and Albania.
>
> **INGEBORG FOSS,** 42, a nurse from the Norwegian Red Cross, came from her first ICRC mission, in Pakistan, and had only arrived in Chechnya days before her murder.
>
> **NANCY MALLOY,** 51, a nurse manager from the Canadian Red Cross, had seen hazardous duty in Ethiopia, Kuwait (for the Federation), Belgrade, Goma (for the Federation) and was with the ICRC when the hospital opened in September.
>
> **GUNNHILD MYKLEBUST,** 55, a nurse also from the Norwegian Red Cross, began this ICRC assignment, which was her first, after serving with Norwegian humanitarian organizations in Israel and Bosnia.
>
> **SHERYL THAYER,** 40, a nurse from New Zealand, had worked for the ICRC in Thailand and for two tours in Afghanistan.
>
> **FERNANDA CALADO,** 49, a Spanish nurse under contract to the ICRC for 12 years, had risked her life previously in Angola, Pakistan, Israel, Cambodia, Somalia, Rwanda, and Kenya.

The head of the Novye Atagi office, Christophe Hensch, who had witnessed the armed robbery weeks before, survived his wounds.

Geneva went into mourning, a repeat of the grief felt earlier, in June, after the murder of three delegates in Burundi. At the memorial service in St. Peter's Cathedral, ICRC President Cornelio Sommaruga condemned the massacre as

> [A]n attack on the very core of what the Red Cross originally was set out to do and still is: the medical mission to prevent and alleviate human suffering.
>
> If we accept that such intolerable behavior will continue in a climate of indifference, then the intolerable will become the rule. This tragedy must be seen from this perspective if it is to have any meaning. It is a challenge to us all: Governments, National Red Cross and Red Crescent Societies, their Federation, the ICRC and other humanitarian organizations alike, all those mourning with us today. There are clear choices to be made, and clear messages to be given. It is by pursuing our efforts, our campaign on behalf of humanity in the very midst of conflict that we shall honour the memory of our colleagues who have given their lives for the humanitarian cause.
>
> Equally it is important that the States Party to the Geneva Conventions reflect on how they will react to this tragedy. *It is an obligation of the Geneva Conventions not only to protect but also to ensure protection. How will governments, individually and collectively, respond?* [My emphasis.] For many, the Red Cross and Red Crescent represents a lifeline—a lifeline in responding to the needs of elderly people who live alone, of disenchanted youth, of the sick and the wounded, of victims of disaster and conflict. With each attack against the emblem and the spirit of the Red Cross and the Red Crescent that lifeline becomes a little weaker. The question that lies unanswered before all of us is "who will take what new action to restore respect for the emblem and the protection of those who work under it?"[4]

The previous day, in meeting the bodies of the delegates at the airport, President Sommaruga stated that these war crimes "lend a new urgency to the ongoing discussions on the establishment of an international criminal court."[5]

The hospital at Novye Atagi, the "haven of hope and humanity," remains open under the direction of the local staff. The other 14 ICRC delegates were evacuated from Chechnya hours after the murders.

Geneva's removal of delegates who are in clear and immediate danger is mandatory. They cannot protect themselves if all the standard techniques of gaining consent from the warring parties—acting with neutrality and impartiality, establishing good personal relations, being discrete, and delivering goods and services—prove insufficient. Painful as it is to pull out, doing so not only protects the lives of delegates, it also makes a statement about those responsible. They are designated brutes and criminals—beyond civilization. So identified, a target is created for the more "political" international community to deal with and to repress. This is the essence of President Sommaruga's message quoted above. *He was shifting missions.* When the main mission is temporarily defeated, as it was at Novye Atagi, the unspoken mission is activated as the ICRC's prime concern. The ICRC: calls for UN and other third party intervention; exchanges key intelligence on the perpetrators (though it never admits it); encourages constraints on warring parties in the war zones and efforts to resolve the conflict; lobbies for external sanctions on the war criminals; feeds the media, directly and indirectly, with information sustaining charges of war crimes; and marshalls support from other humanitarian organizations to restore conditions for the restoration of relief efforts.

If the main mission of relief and protection facilitates the unspoken mission of undermining internal wars, then paradoxically, the unspoken mission can facilitate and restore the main mission when it breaks down. With today's internal wars and all their hazards, the main mission and the unspoken mission are inseparably linked. The unspoken mission cannot be abandoned, for to do so would mean the abandonment of the ICRC's main mission. The unspoken mission protects the main mission by mobilizing allies whose political intervention can protect the ICRC's and others' humanitarian assistance. Hazards must be countered and minimized. Third-party presence in internal wars—the more massive the better—is the prerequisite for the unspoken mission. As President Sommaruga indicated, "It is an obligation of the Geneva Conventions not only to protect, but also to ensure protection. How will governments, individually and collectively, respond?"

THE ICRC AS A TARGET

What was the ICRC doing in Chechnya that made its operation a target of barbarity? The answer is that the ICRC was doing what it usually does, which, of course, is subversive to internal conflicts. The masked gunmen, therefore, represent a group or groups who do not want the conflict to end, who do not want "havens of hope and humanity" just yet.

The ICRC has been present in the Northern Caucasus since July 1993. Originally, it provided assistance to people who had been displaced or detained when violence broke out between the Ossetians and Inguish people. This operation provided a base when the Chechen war erupted the next year.

In September 1994, when the crisis came to a head, the ICRC brought in emergency medical aid and

> launched a formal appeal to the warring parties to respect the basic rules on international humanitarian law. In particular, it urged them to spare civilians and their property, to ensure humane treatment of those who surrendered, captured combatants and civilians arrested in connection with the conflict, to refrain from taking hostages, and to respect the sick and wounded, medical personnel, establish-

> ments and vehicles and the red cross and red crescent emblems protecting them.[6]

These efforts failed to restrain all-out war. The ICRC then shifted to efforts to gain access to detainees and to develop protection and tracing activities for civilians. Medical facilities were rehabilitated when necessary, and 17 hospitals were provided with basic supplies.[7]

Over the course of the 21-month war between the Russian Federation troops and the Chechen separatist fighters, the ICRC disseminated tens of thousands of booklets on IHL to both sides and, using radio and television, urged them to follow humanitarian principles. Hundreds of detainees were visited each year, often in the face of severe resistance. In 1995, the ICRC distributed 190,837 family parcels, delivering 2,753 tons of relief supplies to Grozny alone.[8] The extent of ICRC relief—which for a time was the only international humanitarian organization operating in Grozny—is indicated when one considers that over half the population of that shrunken city of 120,000 depended upon ICRC assistance. An even greater number of displaced persons throughout Chechnya relied upon ICRC parcels to live. Tracing activities reached the point where a peak of 5,000 messages were exchanged per month. Medical supplies eventually reached 30 health facilities.[9]

The restoration of water supplies and sanitation became a main concern. Over 900 kilometers of mains became clogged or destroyed during the war. Soon after the cease-fire, ICRC engineers in Grozny had restored a daily production capacity of 1 million liters of chlorinated water, 400,000 liters of which were transported in tanker trucks to 57 distribution points.[10] The ICRC was the major source of school supplies for students and for "meals on wheels" for the elderly and disabled.

Seventy delegates and 400 local staff had a visible presence in the Northern Caucasus. Most of them worked in and out of Chechnya before the murders. With a modest UNHCR presence, scant other NGO presence, and only a handful of observers from the OSCE, the ICRC became the principle foreign presence in Chechnya. This made it the principle target of those who wanted to clear the area for conflict.

With few exceptions, before the ICRC enters a war zone it negotiates with the political or military authorities who can affect its operations; whether with the government or the rebels, it needs their consent in order to protect its workers. And to get the consent, the ICRC pledges to observe its principles, especially the big three of neutrality, impartiality, and independence. It tries to develop good personal and personnel relations with all sides and all leaders. It avoids judgmental behavior when it can, in spite of dealing with the inevitable unsavory characters. It avoids preachy, self-righteous arguments about how humanitarian its workers are. It relies on law and principles and patience to make its case for serving the victims of war. It does not inflate its role—although some accuse it of doing so—by advertising its exploits in the media.

Warring parties know instinctively that the ICRC or any third party will get in the way of their war to some extent. Yet, their leaders need ICRC's services to feed, heal, trace, clothe, and shelter their own people. In the minds of warriors, the ICRC can be made a de facto ally. Even the ICRC's efforts at conflict resolution are often welcomed because few warring parties want an unconditional, absolute victory. An agreement, as in Guatemala, Bosnia, or Sierra Leone, is perfectly acceptable if it preserves and advances core interests.

Of course, some warring parties want unconditional victory, and that is the greatest hazard to ICRC workers. A sophisticated awareness of the unspoken mission is not necessary for a warring party who wants the war to continue to target the ICRC. It wants a clear field of fire in order to bring it unalloyed triumph. Its leaders want the ICRC out.

Immediately after the hospital murders, most observers concluded that getting foreigners out of Chechnya was the motive for the crime.

The head delegate in the North Caucasus, Fery Aalame, admitted that "It was obviously a very, very dangerous situation, but we could never have expected the ICRC to be targeted like this—I mean there are 100 different ways to give us the message to leave."[11] Daniel Augstburger, an ICRC delegate at the New York UN delegation, told me, as he had told the press, that "Some [people within the warring] parties saw themselves at a disadvantage [with the ICRC presence]. They want to put an end to the peace process, to conduct war without witnesses."[12] "There is a chain of logic which links the killings to other recent events," said Zenon Kuchciak, deputy chief of the OSCE in Grozny. "These acts may indeed be intended to destabilize the political situation . . . by scaring away foreigners."[13]

A Pentagon source who monitors violent groups that could be a problem for the United States said the message of the killings was: "Leave us alone. Don't mess with us."[14] He blamed Islamist extremists, supported from abroad, who want everyone out—Russians, UNHCR, ICRC, and any other foreign group. This would isolate the Chechen moderates who want to carry out the August agreement with the Russians, and who would get tacit Russian support when the internal war resumes. Supporters of an Islamic state, a state that could be linked to other similar regimes and control its oil and gas (as opposed to international investors who would depoliticize the oil and gas), want to hide and isolate their struggle.

This is a plausible scenario. Chechnya is a clannish society, well known for its warrior tradition and for clans that increasingly use religion as a political weapon. As one Russian social scientist describes the internal struggle:

> The essence of the interaction between the long-term and short-term aspects of the conflict between Russia and Chechnya is revealed in the positions of the sides of the internal Chechen conflict. One side stubbornly insists on an isolationist path of returning to their [sic] own historical roots and modernization along the lines of the oil-producing Islamic states, but this would inevitably—at least in the foreseeable future—lead to a marked decline in the level of civilization in Chechnya and to a revival of clan-tribal and tariqat [religious group] relations. The other side wants to keep Chechnya a part of Russia, to sustain the civilization level that has been achieved as a result of union with Russia, and to modernize Chechnya on a European model together with Russian society as a whole.[15]

One can certainly argue with the assertion above that the Chechen moderates still want to be embraced by Russia or maintain a European civilization, but the internal clash of civilizations is clear. The extremists who robbed the ICRC hospital stole medical supplies and communications gear, materials essential to conduct war. Head-strong separatists and Islamic extremists undoubtedly were inflamed by the Christian red cross symbol, the perception of the ICRC delegates as "spies and hypocrites," and by witnessing the ICRC's good relations with Russian authorities and Chechen moderates.

Another scenario, however, became the initial story line from government officials in Moscow and Chechnya. A foreign correspondent in Moscow summarized this broader explanation for the murders a week after the event.

> The situation in Chechnya is still fuzzy, with no sign of it being defuzzed. The Chechen government, separatist leaders who are now in charge, has been blaming [the murders] on Russian secret services, [claims they were] carried out by Chechen "mercenaries and traitors," and [is] calling [them] an attempt to: wreck the peace process (Russian officials say the same); discredit their government; disrupt the January 27 presidential and parliamentary elections; prevent international election observers from coming (seen as important for giving the elections legitimacy); and create an impression that Chechnya is populated by bandits. Many Russians [already] believe this, and there are Chechen organized crime groups operating in Moscow [which supports this view].[16]

Similar views were expressed in Chechnya. "The Red Cross workers were killed for one reason," said Adlan Khasanov, a Chechen journalist: "so that the world would see it and say that we are all bandits to begin with—that it is better just to trust the Russians to deal with us as they see fit."[17] Russian nationalists who do not want Chechnya to separate also have an interest in rekindling the war.

Whatever the political motive, the common denominator is the need for war. Targeting the ICRC confirms that its presence is not congenial to internal wars.

The risks dramatically highlighted by the Chechen massacre brought about an urgent ICRC review of its security policies. All 55 heads of delegations in world conflict areas were recalled to Geneva in late January 1997. Their three-day meeting resulted in the approval of three major departures form previous security policies.

First, in areas where Red Cross workers and facilities are very easy, soft targets for bandits or warring parties, armed guards will be hired for protection. "This could have been done in Chechnya," ICRC spokesman Tony Burgener told me, "where everybody seemed to have a gun."[18] Permission from local authorities will be required, only local forces—police or private—will be hired, and no one will be armed if it would provoke a confrontation.

Second, in cases where the Red Cross flag is seen as a Western or alien symbol, the flag will not be flown. "The Red Cross flag has served us well in general," said Luc Deneys an ICRC legal officer after the meeting. "Historically, the Christian crusaders used a red cross on their uniforms, so in the Middle East it is not liked much."[19] Unwilling to use the red crescent symbol in Muslim areas—which could be interpreted as an admission that the red cross is a religious symbol—the ICRC in "serious situations" will lower its visibility.

Third, other steps will be taken by the ICRC to lower its visibility as a "huge, rich, Western organization—a white elephant." More local citizens will be hired, less use will be made of modern, new vehicles, and more attempts will be made to blend into the local culture.

The ICRC has found it must adjust to the perils of the unspoken mission.

And it needs political/military support.

The ICRC's message to the international community is clear. You endorse war if you stand idly by as humanitarian workers suffer intimidation, kidnapping, and murder. Of course, peace operations by the UN or a regional organization can be hazardous, but why should the ICRC and other humanitarian organizations monopolize the hazards? Identifying the parties who attack the ICRC identifies the war lovers—those who the peacemakers must deter, coerce, bargain with, bring to war-crimes tribunals, or even, in extreme cases, combat in order to bring peace and to further discredit the institution of war.

This, in essence, is the unspoken mission.

Appendices

APPENDIX A: The Committee

All are Swiss citizens and have the authority to define the general policy and principles which guide the activities of the organization.

Mr. Cornelio Sommaruga, President, Doctor of Laws of Zurich University, (member of the Committee since 1986).

Mr. Pierre Keller, Vice-President, Doctor of Philosophy in international relations (Yale), banker, (1984).

Mr. Eric Roethlisberger, permanent Vice-President, Doctor of Political Science of the Graduate Institute of International Studies in Geneva, (1994).

Mr. Ulrich Gaudenz Middendorp, Doctor of Medicine, lecturer at the Faculty of Medicine of Zurich University, former head of the surgical department of the Cantonal Hospital, Winterthur, (1973).

Ms. Renée Guisan, General Secretary of the international "Institut de la Vie," head of medico-social institutions, member of the International Association for Volunteer Effort, (1986).

Mrs. Anne Petitpierre, Doctor of Laws, barrister, Professor at Geneva Law Faculty, (1987).

Mr. Paolo Bernasconi, Bachelor of Laws, barrister, lecturer in economic penal law at the Universities of St. Gallen and Zurich, former Public Prosecutor in Lugano, member of the Swiss *Pro Juventute* Foundation, (1987).

Mrs. Liselotte Kraus-Gurny, Doctor of Laws of Zurich University, (1988).

Ms. Susy Bruschweiler, nurse, former Director of the Swiss Red Cross College of Nursing in Aarau, Chairwoman of S-V Service contract catering, (1988).

Mr. Jacques Forster, Doctor of Economics, Professor at the Institute for Development Studies in Geneva, (1988).

Mr. Jacques Moreillon, Bachelor of Laws, Doctor of Political Science, Secretary General of the World Organization of the Scout Movement, former Director General at the ICRC, (1988).

Mr. Max Daetwyler, graduate in Economics and Social Sciences of the University of Geneva, Scholar in Residence of the International Management Institute (IMI) of Geneva, (1989).

Mr. Rodolphe de Haller, Doctor of Medicine, former lecturer at the Faculty of Medicine of the University of Geneva, President of the Swiss Association against Tuberculosis and Lung Diseases, (1991).

Mr. Daniel Thürer, Master of Laws (Cambridge), Doctor of Laws, Professor at the University of Zurich, (1991).

Ms. Francesca Pometta, Bachelor of Arts, former Swiss Ambassador, (1991).

Mr. Jean-François Aubert, Doctor of Laws, Professor at the University of Neuchatel, former member of the Swiss Parliament, (1993).

Mr. Joseph Feldmann, Doctor of Philosophy, former Professor at the University of St. Gallen, retired Lieutenant General of the Swiss army, (1993).

Mrs. Lilian Uchtenhagen, Doctor of Economics of the University of Basel, former member of the Swiss Parliament, (1993).

Mr. Georges-André Cuendet, Bachelor of Laws of the University of Geneva, graduate of the Institute of Political Studies of the University of Paris (France), Master of Arts of Stanford University (USA), member of the Administrative Council of Cologny (Switzerland), (1993).

Mr. Ernst A. Brugger, Doctor of Natural Science, general manager of FUNDES (a private foundation for sustainable development), consultant for economic development issues, professor at the University of Zurich, (1995).

Source: *ICRC Annual Report 1995,* (Geneva: ICRC, May 1996), pp. 343-44.

APPENDIX B

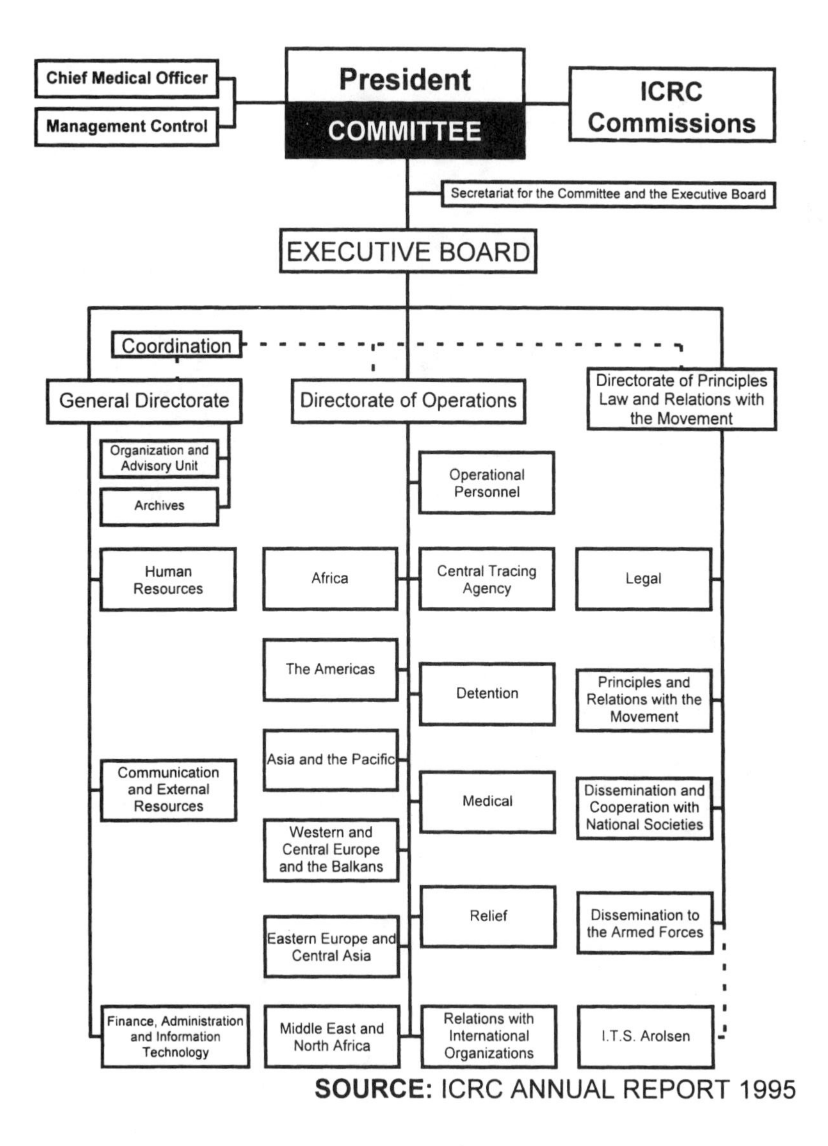

SOURCE: ICRC ANNUAL REPORT 1995

Notes

CHAPTER 1

1. Brian Peachment, *The Red Cross Story: The Life of Henry Dunant and the Founding of the Red Cross* (Oxford: Pergamon Press, 1977).
2. As quoted in Elizabeth Brown Pryor, *Clara Barton: Professional Angel* (Philadelphia: University of Pennsylvania Press, 1987), p. 157.
3. Ian McAllister, *Sustaining Relief with Development: Strategic Issues for the Red Cross and Red Crescent* (Dordrecht, The Netherlands: Martinus Nijhoff Publishers, 1993), pp. 5-6.
4. Georges Willemin and Roger Heacock, *The International Committee of the Red Cross* (Dordrecht, The Netherlands: Martinus Nijhoff Publishers, 1984), p. 21.
5. Pryor, pp. 189-90.
6. *ICRC Annual Report 1995,* (Geneva: ICRC, May 1996), p. 299.
7. Willemin and Heacock, p. 188.
8. *ICRC Annual Report 1994,* (Geveva: ICRC, April 1995), p. 7.
9. Christophe Swinarski, ed., *Studies and Essays on International Humanitarian Law and Red Cross Principles: In Honor of Jean Pictet* (Geneva: Martinus Nijoff Publishers, 1984), p. XLII.
10. Claudio Caratsch, "Humanitarian Design and Political Interference: Red Cross Work in the Post–Cold War Period," *International Relations,* Vol. 11, (April 1993), p. 305.
11. Willemin and Heacock, p. 27.
12. Ibid.
13. *ICRC Annual Report 1995,* p. 20.
14. Ibid.
15. Caratsch, p. 312.
16. *ICRC Annual Report 1994,* p. 14.
17. Caratsch, p. 312.
18. *ICRC Annual Report 1995,* p. 21.
19. Ibid., p. 22.
20. Ibid., p. 25
21. Ibid., p. 31.
22. Ibid., pp. 326-27.

CHAPTER 2

1. Desert Storm or the second Persian Gulf War is the major exception.
2. The 1979-88 Iran-Iraq War is one exception, in which both superpowers' national interests called for a stalemate. At various times the United States and USSR supported both sides. Some analysts would say that the Falklands/Malvinas War between the UK and Argentina is another exception because both belligerents

were on the same Cold War side. But even here, the United States provided intelligence data to the British, and the Soviets gave modest lip service to the Argentines.

3. One major exception was the Nigerian Civil War, which had little Cold War ideological content.
4. "ICRC Report to the 26th International Conference," *International Review of the Red Cross,* No. 311 (Geneva: ICRC, March-April 1996), p. 219.
5. Umesh Palwankar, ed., *Symposium on Humanitarian Action and Peace-keeping Operations: Report* (Geneva: ICRC, 1994), p. 63.
6. Peter Hansen, "Humanitarian Aid on an International Scope," *The Christian Science Monitor,* 15 August 1995, p. 18.
7. ICRC Conference Handout, December 1995, p. 9.
8. Rodolfo A. Windhausen, "UN's Refugee Agency Defies Worn Stereotypes," *The Christian Science Monitor,* 21 December 1995, p. 19.
9. *ICRC Annual Report 1994,* pp. 54-55.
10. Ibid., pp. 146-47.
11. Ibid., p. 38
12. *The New York Times,* 8 November 1995, p. A8.
13. François Jean, *Life, Death and Aid: The Medécins sans Frontières Report on World Crisis Intervention* (London: Routledge, 1993), p. 10.
14. Ibid., p. 129.
15. ICRC Conference Handout, p. 7.
16. Yves Sandoz. "A Time to Speak Out for the Victims." *Crosslines,* Vol. 3(4) (December 1995/January 1996), p. 29.
17. M. Mercier, *Crimes sans Châtement* (Bruxelles: Bruylant, 1994) as cited in the *ICRC Annual Report 1994,* p. 155.
18. As quoted in André Durand, "The Development of the Idea of Peace in the Thinking of Henry Dunant," extract from the *International Review of the Red Cross* (January/February 1986), p. 5.
19. As quoted in Andre Durand, "Gustave Monyier and the Peace Societies," *International Review of the Red Cross,* No. 314 (September/October 1996), pp. 546-47.
20. Lori Fisler Damrosch, ed., *Enforcing Restraint: Collective Intervention in Internal Conflicts* (New York: Council on Foreign Relations Press, 1993).
21. Author's notes from the 26th International Conference of the Red Cross and Red Crescent.
22. Kim Gordon-Bates, "Challenging Inhumanity," *Crosslines,* Vol. 3(4) (December 1995/ January 1996), p. 28.
23. "Civilians in War," handout, 26th International Conference of the Red Cross and Red Crescent, p. 3.
24. Interview with the author, Geneva, 5 December 1995.
25. Sandoz, *Crosslines,* p. 29.
26. Interview with the author, New York, 20 November 1995.
27. Yves Sandoz, "To the Attention of National Societies," (Geneva: ICRC, 29 November 1994), pp. 1-3.

CHAPTER 3

1. The Red Cross' Council of Delegates is made up of representatives of the three constituent units of the Movement: the ICRC, the Federation, and the National Societies.
2. Data provided by ICRC's New York delegation.
3. *Code of Conduct,* p. 3, as distributed by the ICRC. This brochure is the source of all the material on the Code, except as noted.
4. International Federation and the ICRC, "Principles and Response in International Humanitarian Assistance, Commission II: Humanitarian Values and Response to Crisis." (Geneva, 1995), p. 5. (Hereafter, Commission II).
5. Ibid.
6. Interview with the author, Geneva, 5 December 1995.
7. Commission II, p. 15.
8. Ibid.
9. A statement by George Weber, secretary general of the International Federation of the Red Cross and Red Crescent, at the International Conference's final news conference, Geneva, 7 December 1995.
10. As quoted in Larry Minear and Thomas G. Weiss, *Humanitarian Politics,* Headline Series No. 304 (Ithaca: Foreign Policy Association, 1995), p. 14.
11. François Jean, *Life, Death and Aid: The Médecins sans Frontières Report on World Crisis Intervention* (London: Routledge, 1993), p. 10. The MSF has a policy to publicize the world's ten most urgent humanitarian crisis in its annual report. Whistleblowing on ethnic cleansing, starvation, and other war crimes will be analyzed in more detail in the next chapter.
12. Georges Willemin and Roger Heacock, *The International Committee of the Red Cross* (Dordrecht, The Netherlands: Martinus Nijhoff Publishers, 1984), p. 23.
13. *The New York Times,* 13 October 1995, p. A6.
14. As reported in *The Christian Science Monitor,* 27 September 1995, p. 6.
15. Ibid.
16. "Civilians in War," 26th International Conference of the Red Cross and Red Crescent, (Geneva: ICRC, 1995), p. 14.
17. "Humanitarian Values and Response to Crisis," 26th International Conference of the Red Crescent, (Geneva: ICRC, 1995), p. 4.
18. Margaret Mead, "Warfare is Only an Invention—Not a Biological Necessity," reprinted in John A Vasquez, ed., *Classics of International Relations,* 3rd ed. (Upper Saddle River, NJ: Prentice Hall, 1996), p. 222.

CHAPTER 4

1. The phrase "humanitarian politics" is the title of a Headline Series book (New York: Foreign Policy Association, 1995) by Larry Minear and Thomas G. Weiss. Both authors are at Brown University's Thomas J. Watson, Jr. Institute for International Studies. They were among the first to recognize that humanitarian action is political action.
2. Umesh Palwankar, ed., *Symposium on Humanitarian Action and Peace-keeping Operations, A Report* (Geneva: ICRC, 22-24 June 1994), p. 5.

3. Commissions at the International Conference include representatives of the three constituent parts of the Red Cross and Red Crescent Movement and the governments who are signatories to the four 1949 Geneva Conventions. As commissions, they study a broad problem, prepare reports, and present draft resolutions to a full plenary meeting.
4. Palwankar, p. 94. Yves Sandoz is a member of the Executive Board and Director of Principles, Law and Relations with the Movement.
5. Ibid., pp. 94-95.
6. Ibid., pp. 30-31.
7. Ibid., p. 38.
8. ICRC Report, "Meeting of Experts on the Applicability of International Humanitarian Law to United Nations forces," (Geneva, July 1995). Besides Sise, the participants were: M. Andre Andries, Premier Avocat pres la Cour Militaire, Belgique; M. Philippe Chuma, Centre des droits de l'homme, Palais des Nations, Geneve; Professor Dr. Horst Fischer, Academic Director, Institute for International Law of Peace and Armed Conflict, Ruhr-Universitat Bochum, Germany; Professor Christopher Greenwood, Director of Studies in Law, Cambridge, United Kingdom; Lt.-Col. Joseph C. Holland, Office of the Judge Advocate General, Canada; Lt.-Gen. (retd.) Satish Nambiar, India; and from the ICRC: Marco Sasso!li, Deputy Head of the Legal Division; Aleardo Ferretti, Deputy Head Dissemination to the Armed Forces Division; Antoine Bouvier, Member of the Legal Division; and Lydie Ventre, Trainee, Legal Division.
9. 26th International Conference of the Red Cross and Red Crescent, "Principles and Response in International Humanitarian Assistance and Protection: Commission II: Humanitarian Values and Response to Crisis (Item 2 of the provisional agenda)," (Geneva, 1995), p. 4.
10. Ibid., pp. 4-5.
11. Ibid., p. 6.
12. "Commission II, Item 2: Draft Resolution," Rev. 3, (Geneva, 7 December 1995), p. 5.
13. Interview with the author, New York, 20 November 1995.
14. The Conference "Weapons of War, Tools of Peace" was sponsored by the International Center for Humanitarian Reporting and was held December 5-6, 1995, at the Palais des Nations, United Nations, Geneva.
15. Palwankar, pp. 100-101.
16. Ibid., p. 88.
17. Ibid., p. 63
18. Ibid., p. 62.
19. These are the words of Jean de Courten, but all ICRC officials repeat this phrase as equivalent to a religious mantra.
20. "Commission I, Item 3: Protecting the Civilian Population in Periods of Armed Conflict, Draft Resolution," (Geneva, 15 November 1995), p. 2.
21. "Commission II, Item 2, Principles and Actions in International Humanitarian Assistance and Protection, Draft Resolution," Rev. 3 (Geneva, 7 December 1995), p. 5.
22. ICRC Report, "Meeting of Experts on the Applicability of International Humanitarian Law to United Nations Forces," (Geneva, July 1995), pp. 42-43.

23. Ibid., p. 45.
24. Palwankar, p. 42.
25. Ibid., p. 44.
26. "Commission I, Item 3, Draft Resolution," (Geneva, November 15, 1995), p. 4.
27. Ibid., p. 5.
28. "Commission II, Item 2 of the provisional agenda," p. 18.
29. Ibid., pp. 18-19.
30. Ibid., p. 19.
31. Ibid., p. 98.
32. Ibid., p. 89.
33. "Commission I, Item 3, Draft Resolution," (Geneva, 15 November 1995), p. 2.
34. "Commission II, Item 2 of the provisional agenda," p. 18.
35. Palwankar, p. 98.
36. Such as those Institutes at the University of Virginia and Duke University in the United States.
37. While attending the University of Virginia's Institute on National Security Law in 1995, our class was briefed by the head of the White House National Security Council legal staff, and, if my memory serves me, he remarked that his staff had quadrupled in ten years.
38. "Commission I, Item 2 of the provisional agenda," p. 7.
39. Ibid., pp. 25, 27.

CHAPTER 5

1. *ICRC Annual Report 1994,* (Geneva: ICRC, April 1995), p. 148.
2. Ibid.
3. Ibid., p. 149.
4. Ibid., p. 148, and reiterated in *ICRC Annual Report 1995,* (Geneva: ICRC, May 1996), p. 164.
5. *ICRC Annual Report 1994,* pp. 146-48, and *ICRC Annual Report 1995,* pp. 170-71.
6. ICRC *Saving Lives,* (Geneva: ICRC, 1995), p. 8. The title indicates the main task of the ICRC in Bosnia-Herzegovina.
7. Ibid., p. 9.
8. *ICRC Annual Report 1995,* p. 172.
9. Nik Gowing, "Real-Time Television Coverage of Armed Conflicts and Diplomatic Crises: Does It Pressure or Distort Foreign Policy Decisions?" Working Paper 94-1 (Cambridge: John F. Kennedy School of Government, June 1994), p. 44.
10. *Saving Lives,* p. 3.
11. Ibid., p. 5.
12. Ibid., p. 11.
13. Ibid., p. 12.
14. Ibid., p. 5.
15. ICRC, *Challenges of the Nineties,* (Geneva, 1995), p. 26.
16. Ibid., p. 8.
17. *ICRC Annual Report 1995,* pp. 169, 173.
18. Ibid., p. 165.

19. *ICRC Annual Report 1994,* p. 53.
20. Interview with the author, the ICRC UN mission in New York, 20 November 1995.
21. *ICRC Annual Report 1994,* pp. 34-35.
22. Ibid., p. 58.
23. Ibid., pp. 53, 56.
24. Ibid., p. 60.
25. *Challenges of the Nineties,* pp. 8-9.
26. Ibid., with updated figures provided by the ICRC delegation at the UN.
27. *ICRC Annual Report 1994,* p. 57.
28. Ibid., p. 55.
29. *ICRC Annual Report 1995,* p. 58.
30. Ibid., pp. 59-60.
31. *The New York Times,* 20 November 1996, p. A10.
32. Gowing, "Real-Time Television Coverage," p. 17.
33. Bill Berkeley, "The Longest War in the World," *The New York Times Magazine,* 3 March 1996, p. 59.
34. *ICRC Annual Report 1994,* p. 91.
35. Ibid, p. 90.
36. *The New York Times,* 5 February 1996, p. A8.
37. Ibid.
38. *The Christian Science Monitor,* 27 September 1995, p. 7.
39. *ICRC Annual Report 1994,* p. 103.
40. *The New York Times.* 5 February 1996, p. A8.
41. *ICRC News,* No. 42 (19 October 1995), p. l.
42. *ICRC Annual Report 1995,* p. 123.
43. *The New York Times,* 4 October 1996, p. A3.
44. *The Christian Science Monitor,* 27 September 1995, p. 7.
45. Ibid.
46. *The New York Times,* 5 February 1996, p. A8.
47. Ibid.
48. *The New York Times,* 4 October 1996, p. A3.

CHAPTER 6

1. *The New York Times,* 30 December 1996, p. A4.
2. *The New York Times,* 20 September 1996, p. A16.
3. Richard H. Immerman, *The CIA in Guatemala: The Foreign Policy of Intervention* (Austin: University of Texas Press, 1985), p. 28.
4. Walter LaFeber, *Inevitable Revolutions: The United States in Central America,* 2d ed. (New York: W.W. Norton, 1993), p. 120.
5. Immerman, p. 81.
6. LaFeber, p. 124.
7. Ingrid Flory and Alex Roberto Hybel, "To Intervene or Not To Intervene: An Analysis of U.S. Actions Toward Guatemala and Bolivia in the Early 1950s," *The Journal of Conflict Studies,* 15, no. 2, (Fall 1995), pp. 82-83.

8. Dwight D. Eisenhower, *Mandate for Change, 1953-56* (Garden City, NY: Doubleday, 1963), p. 421.
9. Ibid., p. 422.
10. Ibid, p. 423.
11. Ibid.
12. Immerman, p. 155.
13. Eisenhower, p. 427.
14. LaFeber, p. 172.
15. Suzanne Jonas, "Dangerous Liaisons: The U.S. in Guatemala," *Foreign Policy,* 103 (summer 1996), p. 147.
16. Human Rights Watch/Americas Report, "Guatemala—Return to Violence: Refugees, Civil Patrollers, and Impunity," (New York: Human Rights Watch, January 1996), p. 5.
17. Immerman, p. 197.
18. Rachel M. McCleary, "Guatemala: Expectations for Peace," *Current History,* 95, no. 598 (February 1996), p. 89.
19. Jonas, p. 158.
20. Human Rights Watch/Americas, pp. 8-11.
21. Ibid.
22. *The Christian Science Monitor,* 16 May 1996, p. 23.
23. *The New York Times,* 1 April 1996, p. A ll.
24. Ibid.
25. *The New York Times,* 24 March 1996, p. 6.
26. Jonas, p. 155.
27. *The Washington Post,* 28 June 1996, p. l.
28. *The Guatemala News* (15 March 1996, p. 2) printed her internet message that urged Guatemalan supporters to petition the White House to declassify documents relating to the murders and to "cut off visas to known military human rights violators."
29. *The New York Times,* 24 March 1996, p. 6.
30. Jonas, p. 150.
31. *ICRC Annual Report 1994,* Geneva, p. 268.
32. For example, Hans-Peter Gasser, *International Humanitarian Law: An Introduction,* separate print from Hans Haug, *Humanity for All.* International Red Cross and Red Crescent Movement, Henry Dunant Institute (Berne: Haupt, 1993), p. 69.
33. Marion Harroff-Tavel, "Action Taken by the International Committee of the Red Cross in Situations of Internal Violence," *International Review of the Red Cross,* no. 294 (May-June 1993), pp. 202-03.
34. Gasser, p. 75.
35. Ibid., p. 80.
36. Donald Dochard, ed. *Law of War* (Geneva: ICRC, August 1995), p. 6.
37. McCleary, p. 90.
38. Ibid. She is the author of *I, Rigoberta Menchú—An Indian Woman in Guatemala* (UK: Verso, 1985; distributed by Routledge, Chapman, & Hall, 1985).

39. Jonas, p. 151.
40. Ibid., p. 153.
41. *The New York Times,* 27 March 1996, p. A 8.
42, Ibid.
43. Ibid.
44. I witnessed such a protest by landless Mayan agricultural workers outside the National Palace in March 1996. The police and military looked on at a distance with apparent unconcern.
45. *The New York Times,* 27 March 1996, p. A 8.
46. *The New York Times,* 30 December 1996, p. A4.
47. McCleary, p. 92.

CHAPTER 7

1. President Sommaruga's address is reprinted in the *International Review of the Red Cross,* no. 310 (January-February 1996), p. 20.
2. Ibid., p. 34.
3. ICRC Report, "Challenges of the Nineties," (Geneva, November 1995), p. 21.
4. Ibid., pp. 2-3.
5. Beatrice Megevand, "ICRC Action in Mexico," *International Review of the Red Cross,* no. 304 (January-February 1995), p. 105.
6. Yves Sandoz, "To the Attention of National Societies," a cover letter to his San Remo address entitled "Conflict Prevention and Promotion of International Humanitarian Law;" (Geneva, 29 November 1994), p. 2.
7. Ibid., pp. 1-2.
8. Ibid., p. 6.
9. *International Review of the Red Cross,* no. 310, January-February 1996, p. 34.
10. Ibid., pp. 14-15.
11. Ibid., p. 161.
12. ICRC Report, "Humanitarian Values and Response to Crisis," (Geneva, 1995), p. 12.
13. Sandoz, p. 5.
14. Ibid., p. 6.
15. *The Philadelphia Inquirer,* 5 June 1996, p. 2.
16. *International Review of the Red Cross,* no. 310, p. 31.
17. Ibid.
18. Ibid.
19. *International Review of the Red Cross,* no. 304 (January-February 1995), p. 4.
20. Ibid., p. 30.
21. Ibid., p. 21.
22. Ibid., p. 37.
23. As reprinted in *International Review of the Red Cross,* no. 310, p. 59.
24. *The Christian Science Monitor,* 14 May 1996, p. 7.
25. *The Philadelphia Inquirer,* 18 June 1996, p. A5.
26. *International Review of the Red Cross,* no. 310, p. 22.
27. ICRC Report, "Humanitarian Values and Response to Crisis," p. 112.

28. Ibid.
29. Cornelia Sommaruga, "Strengthening the Coordination of Humanitarian Assistance," *International Review of the Red Cross,* no. 304 (January-February 1995), p. 85.
30. Sandoz, p. 3.
31. Ibid.
32. *ICRC Annual Report 1995,* (Geneva: ICRC, May 1996), p. 56.
33. *The Christian Science Monitor,* 14 June 1996, p. 2.
34. ICRC delegates in the field tend to be young and unmarried. Some delegates complain that they are discarded around age forty or that their jobs are incompatible with families because few war zones are safe enough to take the spouse and children with them. Most delegates, however, see relative youth and lack of family attachments positively—as supportive of independence and risktaking. ICRC lore is full of stories in which delegates' lives hang by a thread, awaiting the action of some bloodthirsty warlord or drunken soldiers. Too often people wearing the red and white emblem have died as a result. The rules of the Geneva Conventions and Additional Protocols prescribe that Red Cross workers are to be inviolable. International law, therefore, should not only guide the behavior of warring parties but also protect Red Cross personnel and equipment. No wonder the Red Cross and Red Crescent Movement values international law and its enforcement so highly. Law facilitates all that it does.

CHAPTER 8

1. *The New York Times,* 19 December 1996, p. A1.
2. Ibid., p. A10.
3. The following information is from the ICRC website [http://www.icrc.ch], which files the ICRC news releases by geographical area and event.
4. Ibid., 19 December 1996, News Release.
5. Ibid., 18 December 1996, News Release.
6. *ICRC Annual Report 1994,* (Geneva: ICRC, April 1995), p. 173.
7. Ibid., p. 176.
8. *ICRC Annual Report 1995,* (Geneva: ICRC, May 1996), p. 201.
9. Ibid., p. 206.
10. ICRC website, Report, "Republic of Chechnya," December 1996.
11. *The New York Times,* 19 December 1996, p. A10.
12. Telephone interview with the author, 20 December 1996.
13. *The Christian Science Monitor,* 20 December 1996, p. 6.
14. Telephone interview on December 22, 1996 by a source who does not want to be identified.
15. V. S. Drozdov, "Chechnya: Exit from the Labyrinth of Conflict," *Russian Social Science Review,* Vol. 37, No. 6 (November/December 1996), pp. 34-35.
16. E-mail from a confidential source.
17. *The Christian Science Monitor,* 20 December 1996, p. 6.
18. Telephone interview with Tony Burgener on 28 January, 1997.
19. *Philadelphia Inquirer,* 25 January, 1997, p. A12.

Index

Y

Z